Knowledge Management: Learning from Knowledge Engineering

Jay Liebowitz

CRC Press
Taylor & Francis Group
Boca Raton London New York

CRC Press is an imprint of the
Taylor & Francis Group, an **informa** business

Published 2001 by CRC Press
Taylor & Francis Group
6000 Broken Sound Parkway NW, Suite 300
Boca Raton, FL 33487-2742

© 2001 by Taylor & Francis Group, LLC
CRC Press is an imprint of Taylor & Francis Group, an Informa business

First issued in paperback 2019

No claim to original U.S. Government works

ISBN-13: 978-0-367-45531-6 (pbk)
ISBN-13: 978-0-8493-1024-9 (hbk)

**Visit the Taylor & Francis Web site at
http://www.taylorandfrancis.com**

**and the CRC Press Web site at
http://www.crcpress.com**

Library of Congress Cataloging-in-Publication Data

Liebowitz, Jay, 1957–
    Knowledge management: learning from knowledge engineering / Jay Liebowitz
        p. cm.
    Includes bibliographical references and index.
    ISBN 0-8493-1024-5
    1. Expert systems (Computer science). 2. Knowledge management. I. Title.
QA76.76.E95 L55 2001
658.4′038—dc21   2001000790
                                                                          CIP

Preface

Who should read this book and why?

I believe this is the first book that addresses exclusively the possible synergies between the disciplines of knowledge engineering and knowledge management. Knowledge management teaches us to learn from others and to share with and nurture others and not to reinvent the wheel. Unfortunately, the knowledge management community is in a paradoxical state in that we are not learning from other disciplines, most notably the knowledge engineering field. Expert and knowledge-based systems methodologies, techniques, and tools can greatly improve the current state of the art in knowledge management. This realization needs to be disseminated and permeated throughout the knowledge management community, and, thus, this book will hopefully serve that function.

Any knowledge manager, chief knowledge officer, director of intellectual capital, knowledge analyst, knowledge engineer, or knowledge management/intelligent systems user should be very interested in reading this book. A student taking a knowledge management or even a knowledge engineering/AI/expert systems/decision support systems course should also find this book very useful. The book gives concise, easy-to-follow, practical information and insights based upon the author's years of experience in the expert systems and knowledge management fields. Any practitioner interested in knowledge management should find this book helpful. It was written to allow one to easily read it while on a plane, at a beach, in bed, or at a desk.

In writing this book, I would like to thank my students of the past 18 years who have taken expert systems/AI courses with me, as well as those students of the last four years who have enrolled in my knowledge management course and shared their insights with me. I would also like to thank my colleagues at the University of Maryland–Baltimore County (UMBC), George Washington University, U.S. Army War College, and James Madison University who helped shape my vision of the "knowledge age." Most importantly, I want to thank my family — Janet, Jason, Kenny — and my parents for letting me pursue my thirst for knowledge and for all their loving support.

Enjoy!

Jay Liebowitz, D.Sc.

Author's bio

Dr. Jay Liebowitz is the Robert W. Deutsch Distinguished Professor of Information Systems at the University of Maryland–Baltimore County (UMBC) in Catonsville, Maryland. He was previously Professor of Management Science in the School of Business and Public Management at George Washington University and has served as the Chair of Artificial Intelligence at the U.S. Army War College. He is the founder and Editor-in-Chief of *Expert Systems with Applications: An International Journal* (published by Elsevier) and *Failure and Lessons Learned in Information Technology Management: An International Journal*. He is the founder and chair of the World Congress on Expert Systems, where typically 40 to 45 countries are represented. Dr. Liebowitz was selected as the Computer Educator of the Year by the International Association for Computer Information Systems. He was a Fulbright Scholar in Canada, the IEEE-USA Federal Communications Commission Executive Fellow, and holds a number of other honors. He has published 28 books and over 220 papers dealing with expert systems, knowledge management, and information systems management. He can be reached at liebowit@umbc.edu.

Contents

chapter one

Knowledge management and knowledge engineering: working together

A little knowledge goes a long way! The following anecdote typifies the value of knowledge:

> A man goes into a New York City bank and asks for a $2000 loan to take a two-week trip to Europe. The loan officer asks the man what collateral he has. The man points to his Rolls Royce parked in front of the bank and gives his car keys to the loan officer. The man gets the loan and comes back two weeks later after returning from Europe to pay back the loan. He asks the loan officer how much he owes. The loan officer replies it will be $2000 for the principal plus $15.46 for interest. "And by the way," the loan officer continues, "I checked your background and noticed you are a multi-millionaire. Why did you need a loan for such a small amount of money?" The man replies, "Where else can I park in New York City for two weeks for only $15.46?!"

The man in this story applied his "knowledge" — his capability to act. He certainly understood the value of knowledge!

Like the man in this story, organizations are beginning to realize that their competitive edge is their employees' knowledge — the intellectual capital of the organization. One Navy captain who serves as the commanding officer of a Navy research lab stated that their main product is knowledge. Similarly, organizations worldwide have realized that their future growth is predicated on how well they create, manage, share, and leverage their knowledge internally and externally. This has created the knowledge management or knowledge sharing movement.

Knowledge management (KM)

Knowledge management is the process of creating value from an organization's intangible assets. Intangible assets, also referred to as intellectual capital, include human capital, structural capital, and customer or relationship capital. Human capital is the brain power — the people knowledge — in the organization. Structural capital refers to intellectual assets that cannot be easily taken home with the employees, such as patents, trademarks, certain databases, and other related items. Customer or relationship capital is what can be learned from the organization's customers or stakeholders. For example, many years ago Johnson & Johnson began to include Italian powder for preventing skin irritation with the medical plasters that it sold. Customers contacted Johnson & Johnson and expressed their enthusiasm for the powder, which evolved into Johnson & Johnson's baby powder, which at one time accounted for 44% of Johnson & Johnson's revenues. In this sense, Johnson & Johnson learned from its customers through the customer demand which generated a new product — baby powder.

Organizations are embracing knowledge management for several reasons. One primary reason is to increase innovation within the firm. Other major factors for engaging in knowledge management include knowledge retention, people retention, and return on vision. By capturing key knowledge before experts retire or leave the firm, knowledge retention can be increased for building the institutional memory or knowledge base via knowledge management efforts. Communities of practice in which people have shared trusts, beliefs, and values are components of knowledge management programs that give people a sense of belonging and allow lessons learned to be shared. Thus, people retention should be increased because employee morale will be enhanced through collaboration and bonding among those communities of practice. Return on vision vs. return on investment should appear through knowledge management efforts, as knowledge management should promote the vision of the organization. It is sometimes hard to quantify a return on investment for knowledge management efforts, so some organizations are encouraging a return on vision approach. Appendix A shows a knowledge management strategy developed for the Federal Communications Commission.

Is knowledge management new?

Sir Francis Bacon coined the expression, "knowledge is power." For knowledge management, the focus is "sharing knowledge is power." With intranets and Web-based technologies, we now have the connectivity to bridge across isolated islands of knowledge. Sharing in a collaborative way to stimulate new ideas is not a new concept, although facilitating this sharing in an electronic networking fashion is somewhat novel. However, the basic underpinnings of knowledge management are not at all new. The basic principles deal with people, culture, and technology. Many experts feel that about 80% of knowledge management involves the people and culture components, and about 20% deals with the knowledge management technologies. The aspects of people and culture are rooted in organizational behavior, human resources management, and fundamentals of management. The technology component of the triad has its foundation in artificial intelligence, knowledge engineering, information technology, library science, and information systems. The real paradigm shift that makes knowledge management difficult is the migration from an individualist, competitive, "knowledge is power" attitude to a collaborative, "sharing knowledge is power" viewpoint. A cultural shift needs to be created to encourage knowledge sharing. In fact, some organizations such as The World Bank and Andersen Consulting have established learning and knowledge sharing proficiencies as part of their annual employee performance evaluations. From elementary school through college, we have been educated and evaluated on individual performance — individualized tests, individual homework assignments, etc. — as opposed to group, collaborative, team sharing performance. This has stymied a knowledge-sharing culture, but new programs encouraging integrated multi-disciplinary programs and emphasizing team problem-solving are being created that promote building a supportive culture for knowledge sharing. In the coming years, these integrated programs will create classes of scientists, engineers, technologists, humanists, and others who will espouse a knowledge-sharing culture.

Knowledge engineering (KE)

Knowledge management is strongly rooted in a discipline called knowledge engineering, a field that involves the development of knowledge-based or expert systems. Knowledge engineering emerged in the 1960s and 1970s, and gained commercial interest starting in the early 1980s. Knowledge engineering grew out of the field of artificial intelligence and focused on building computer programs that simulated the behavior of experts in well-defined domains of knowledge.

The knowledge engineering process involves the capture, representation, encoding, and testing/evaluation of expert knowledge. As such, a knowledge base is built containing the set of facts and heuristics (rules of thumb) relating to the expert's well-defined task of knowledge.

The two fields of knowledge management and knowledge engineering often overlap, as this book emphasizes throughout the following chapters. The knowledge management cycle can be depicted as shown in Figure 1.1. Figure 1.2 shows the KE and KM models when both processes are put together.

Figure 1.1   Knowledge management life cycle.

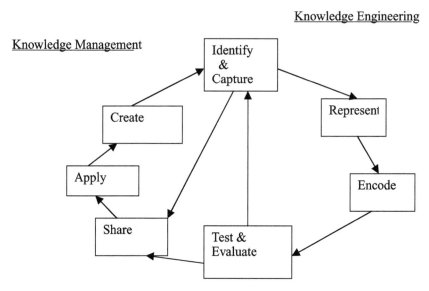

Figure 1.2   Combining the knowledge engineering and knowledge management life cycles.

Knowledge management first involves identifying (or locating) and capturing knowledge. Once the knowledge is captured (including tacit knowledge, which deals with what is in the heads of individuals, and explicit knowledge, which can be easily codified), the knowledge can be shared with others. Then, individuals will apply this shared knowledge and internalize it using their own perspectives. This may produce new knowledge, which then needs to be captured, and the cycle starts over again.

Knowledge engineering has a comparable life cycle to knowledge management. As in knowledge management, knowledge engineering involves both tacit and explicit knowledge, and the emphasis is on capturing tacit knowledge. In knowledge engineering, expertise has to be first located and then captured. After knowledge is acquired and elicited, the knowledge must be represented in rules, cases, or other types of knowledge representation methods. This is similar to developing a KM taxonomy or ontology (in KE terms) in which the KM or KE system will be structured. After the knowledge is represented, it must be encoded into software and then evaluated. Knowledge refinement will probably need to be conducted whereby omitted knowledge needs to be included and captured, and the KE cycle begins again.

What is ahead in this book?

From the brief descriptions in this chapter, one can begin to see where knowledge management and knowledge engineering overlap. Unfortunately, many KM specialists have failed to recognize that KE methodologies, techniques, and tools can greatly enhance the current state of the art of KM. In the following chapters, we will describe how KE and artificial intelligence (AI) techniques are akin to KM and how one can use these processes to improve KM.

chapter two

Knowledge mapping and knowledge acquisition

Developing a knowledge map of an organization is a critical component of knowledge management. This is typically part of the knowledge audit step that attempts to identify stores, sinks, and constraints dealing with knowledge in a targeted business area, and then identifies what knowledge is missing and available, who has the knowledge, and how that knowledge is used. A knowledge map will then be drawn to depict those relationships in that organization.

The methodology used to determine the available and missing knowledge and to capture this knowledge can be borrowed from the knowledge engineering discipline, specifically the knowledge acquisition field. This chapter discusses the knowledge mapping process and how knowledge acquisition techniques can be applied to enhance this process.

Knowledge mapping

According to Jan Lanzing (1997) at the University of Twente, concept mapping is a technique for representing knowledge in graphs. Knowledge graphs are networks of concepts. Networks consist of nodes (points/vertices) and links (arcs/edges). Nodes represent concepts and links represent relations between concepts. Lanzing further explains that concepts and, sometimes, links are labeled. Links can be non-, uni-, or bi-directional. Concepts and links may be categorized; they can be simply associative, specified, or divided in categories such as causal and temporal relations. Research by McDonald and Stevenson (1999) shows that navigation was best with a spatial map, whereas learning was best with a conceptual map.

Typically, concept mapping is performed for several purposes (Lanzing, 1997):

- To generate ideas (brainstorming, etc.)
- To design a complex structure (long texts, hypermedia, large web sites, etc.)
- To communicate complex ideas

- To aid learning by explicitly integrating new and old knowledge
- To assess understanding or diagnose misunderstanding

The idea of knowledge maps and knowledge mapping in the knowledge management field is analogous to the use of concept maps and conceptual mapping. According to Wright (1993), a knowledge map is an interactive, open system for dialogue that defines, organizes, and builds on the intuitive, structured, and procedural knowledge used to explore and solve problems. Knowledge mapping is an active technique for making contextual knowledge representable, explicit, and transferable to others. In knowledge management terms, knowledge mapping relates to conceptual mapping in a very direct way. Specifically, the objective of knowledge mapping is to develop a network structure that represents concepts and their associated relationships in order to identify existing knowledge in the organization (in a well-defined area) and determine where the gaps are in the organization's knowledge base as it evolves into a learning organization.

At Texas Christian University, a knowledge mapping system has been developed by Newbern and Dansereau (1993). They view a knowledge map as a two-dimensional diagram which conveys multiple relationships between concepts using nodes, links, and spatial configuration. The link labels for the knowledge maps are:

	NAME	SYMBOL
Descriptive relationships:	CHARACTERISTIC	C→
	PART	P→
	TYPE	T→
Dynamic relationships:	INFLUENCES	I→
	LEADS TO	L→
	NEXT	N→
Instructional Relationships:	ANALOGY	A→
	SIDE REMARK	S→
	EXAMPLE	E→

Spatial configurations of the knowledge maps may be in hierarchies, clusters, or chains. Knowledge maps can take the form of overview maps, detail maps, and summary maps. Newbern and Dansereau (1993) developed a relationship-guided search to generate a knowledge map:

1. Make a list of important concepts or main ideas. Save this concept list.
2. Pick one concept as a starting node for the map. Put the node in a central location on the paper.
3. Ask the following questions and draw the links on the map. Be sure to label the links before you move on.
   - Can this node be broken down into different types? (Descriptive link label = T)

- What are the characteristics of each type? (Static link label = C)
- What are the important parts of each type? (Static link label = P)
- What are the characteristics of each part? (Static link label = C)
- What led to the starting node? (Dynamic link label = L)
- What does the starting node lead to? (Dynamic link label = L)
- Which things influence the starting node? (Dynamic link label = I)
- What does the starting node influence? (Dynamic link label = I)
- What happens next, or what does this lead to? (Dynamic link label = N/L) Elaborate the map by using analogy links or example links.
4. Pick a new node from the list to start a new map by repeating step 3.
5. As you end the session, review the maps and include any instructional nodes (side comments, definitions, or analogies).

These nodes and linkage connections in a knowledge map may be analogous to the activation of those same ideas in the expert's memory (Wiig, 1993).

Types of knowledge maps

There are no standards or uniformity of how to create a knowledge map. One type of knowledge map, sometimes called organizational map, links people interactions by departments in the organization. Such a knowledge map may resemble the structure below (where X represents a person in the organization and the lines represent frequent interactions):

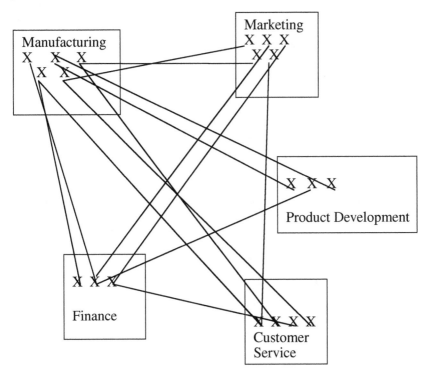

This knowledge/organizational map might suggest that the marketing department is not in frequent contact with product development, which may be problematic as marketing may not know the new products or new versions of current products that are being planned. Marketing should usually be kept abreast of R & D or product development initiatives in order to keep their sales representatives well-informed when meeting with customers/clients.

Another type of knowledge map may link expertise or knowledge areas to experts or individuals. For example, the following could be such a knowledge map:

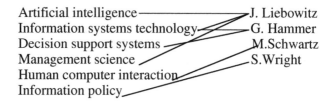

A third type of knowledge map may specifically relate knowledge areas that are available and those that are needed/missing in the organization. One example of this type of knowledge map is:

Critical knowledge (K) Areas	Available K Areas	Missing K Areas	Who has the K
Knowledge management:			
• Human resources management	x		P. Smith
• Strategic planning		x	
• Web-based technologies		x	should be outsourced
• Team building skills	x		

A last type of knowledge map that is frequently used is a type of semantic network with nodes (knowledge areas) and links (relationships between the nodes). One example of this type of partial knowledge map is:

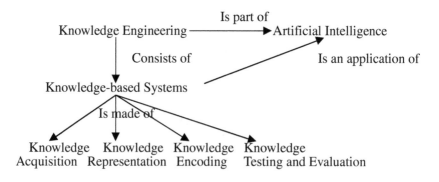

How knowledge acquisition techniques can help build a knowledge map and capture expertise in the KM system

Knowledge acquisition is a critical part of knowledge engineering and knowledge management processes. The main focus is applying techniques to capture expertise and knowledge. Eliciting the set of facts and rules of thumb from experiential learning is one of the key bottlenecks in the knowledge engineering process. Some reasons for this difficulty include:

- What the expert assumes to be common sense may not be common sense to others.
- The knowledge engineering paradox — the more expert an individual, the more compiled the knowledge, and the harder it is to extract or elicit this knowledge.
- The knowledge engineer may misinterpret what the expert is saying.
- Human biases in judgment on the part of the expert and knowledge engineer may interfere with the knowledge being acquired for the knowledge base.
- A single knowledge elicitation session could result in many pages of knowledge elicitation transcripts, resulting in difficulty in organizing the knowledge acquired.

To facilitate the knowledge acquisition process, several techniques could be applied. These include:

- Structured or unstructured interviewing: under structured interviewing, a series of questions is planned in advance and posed to the experts. Asking familiar scenarios (method of familiar tasks) or time- or information-constrained scenarios (constrained processing methods) may help elicit the knowledge from the expert. The "tough cases" method is also used to elicit infrequently occurring cases (special cases) that should be part of the knowledge base. Unstructured interviewing involves an ad hoc approach to knowledge acquisition whereby questions are asked "on the fly" as the expert reasons through various scenarios.
- Protocol analysis: protocol analysis represents a verbal walkthrough whereby the expert talks aloud as he/she reasons through his/her decision making process, highlighting the possibilities and associated logic as to whether a particular possibility was pursued.
- Observation or simulation: the knowledge engineer can observe the expert performing a task (perhaps in a teaching situation) or use simulations to help determine "what if" scenarios for the expert.
- Computer-aided knowledge engineering (CAKE) tools: the expert can directly use a CAKE tool to create decision trees, decision tables/rules, influence diagrams, and the like to provide an intermediate representation for structuring the knowledge base.

To help capture one's tacit knowledge for the knowledge management system, the following generic process (borrowed from the knowledge engineering field) may be used in the knowledge acquisition sessions with the expert:

1. Identify the knowledge engineer, near-expert, and domain expert for the particular domain. The knowledge engineer will be used principally for eliciting knowledge from the expert. The near-expert will be part of the knowledge acquisition team and will be helpful because of his/her familiarity with the domain. The domain expert will be used to capture his/her knowledge and rules of thumb (heuristics) associated with the particular problem area relating to the domain.

2. The first knowledge acquisition session should consist of the domain expert giving a one hour tutorial/introduction (using one of the knowledge acquisition techniques previously highlighted) to the knowledge engineer and near-expert relating to the problem area at hand. This will help the knowledge engineer become more familiar with the problem domain, and this session will also help narrow the scope of the domain. Audio/video-taping and possible transcription from this tutorial should be encouraged.

3. Based upon this tutorial, the knowledge engineer and near-expert could ask the expert in the second knowledge acquisition session the following set of generic questions/instructions, as a starting point:

   - In reviewing your problem area, kindly walk through your reasoning process/decision flow, citing the steps you are following in resolving the problem.
   - What are the factors that influence your decision in resolving this problem?
   - How would you prioritize the importance of these factors in determining your decision?
   - How do these factors contribute to your determination? In other words, what specific conclusions (intermediate or otherwise) can be reached based upon a particular factor and/or a set of factors?
   - What level of certainty (on a scale of one to ten, ten being high) do you associate with each factor in influencing your decision?
   - What shortcuts or helpful techniques have you learned over your years of experience that make you more expert than others (at least as compared to novices)?
   - What are the lessons learned from your possible mistakes or from others in resolving problems in your area of interest?
   - Kindly assemble a set of cases that would include a prioritized list of symptoms/problems and a corresponding (associated) list of prioritized possible diagnoses and decisions.

   Audiotaping and transcription of this session should be encouraged.

4. During the second knowledge acquisition session or starting with the third knowledge acquisition session, the expert should be asked to try to structure his/her knowledge using the following generic template:

   Variables Used in the Decision Process

   Range of Values

   Implications

   What to Do Next

   Sources of Information

   By completing the template, the expert will be developing cause-and-effect relationships. This will also impose some structure to the interviewing process and will help structure/represent the knowledge in the knowledge base. Audiotaping and transcription of this session should be encouraged.

5. Follow-up questions based upon the answers to the questions in the second knowledge acquisition session and this template would be asked by the knowledge engineering team. Additionally, sources of information identified by the expert in completing the template should be acquired for eventual encoding.

   The expert may also want to use some computer-aided knowledge acquisition techniques/tools to help construct decision trees or influence diagrams relating to his/her knowledge. Tools such as RuleBook, by Exsys, Inc., can be used to create the decision trees that get automatically encoded into the Exsys expert system shell. This would facilitate the process of knowledge acquisition, representation, and encoding.

6. Once acquired, the knowledge should next be represented and encoded. Sessions with the expert and the knowledge engineering team will reveal areas where the knowledge base needs to be refined. Then, iterative development of the system (going back through the steps) to reacquire the necessary knowledge should be conducted.

References

Lanzing, J. (1997), *The Concept Mapping Homepage*, University of Twente, the Netherlands, www.to.utwente.nl/user/ism/lanzing/cm_home.htm.

McDonald, S. and R. Stevenson (1999), "Spatial Versus Conceptual Maps as Learning Tools in Hypertext," *Journal of Educational Multimedia and Hypermedia*, Vol. 8, No. 1.

Newbern, D. and D. Dansereau (1993), "Knowledge Maps for Knowledge Management," in *Knowledge Management Methods*, Ed. K. Wiig, Schema Press, Arlington, TX.

Wiig, K., Editor (1993), *Knowledge Management Methods*, Schema Press, Arlington, TX.

Wright, R. (1993), "An Approach to Knowledge Acquisition, Transfer, and Application in Landscape Architecture," University of Toronto, Canada, www.clr.toronto.edu/PAPERS/kmap.html.

chapter three

Knowledge taxonomy vs. knowledge ontology and representation

The 1999 Expert System Workshop on Using Artificial Intelligence (AI) to Enable Knowledge Management (KM) performed a SWOT (strengths–weaknesses–opportunities–threats) analysis on AI in relation to knowledge management (www.bcs-sges.org/kmreport/swot.htm). The analysis showed the following:

Strengths:

- Good support for the identification and acquisition of knowledge through both knowledge elicitation techniques and machine learning techniques
- Experience of categorization through the use of ontologies
- Knowledge modeling and representation techniques
- Understanding of problem solving methods
- Support for sharing and reuse of knowledge
- Experience of the reuse and application of knowledge (e.g., via knowledge-based systems and case-based reasoning)

Weaknesses:

- Poor understanding of cultural or organizational issues
- No support for valuing knowledge assets
- Limited contribution for dealing with tacit knowledge
- AI techniques may discourage a holistic view of knowledge.
- Limited support for recognizing redundant knowledge
- AI systems can be complicated, requiring specialist skills, and can be expensive (AI is viewed as rocket science).

Opportunities:

- AI has a lot of practical experience and techniques to offer KM.
- AI could become better accepted within companies.
- KM is a good vehicle for research funding.

Threats:

- KM is yet another management buzzword.
- KM could/will be overtaken by a new hyped movement in five years, and if AI is too closely associated, it will be perceived as old hat and suffer from (another) backlash.
- KM may be taken over by document management, groupware, and intranet software companies.
- Conversely, knowledge-based systems applications are not KM.

The above analysis demonstrates that AI's strengths can clearly help advance the current state of the art in knowledge management. That is the approach of this book. Even though AI and knowledge engineering techniques are not necessarily considered the saviors of knowledge management, there are overlapping areas of interest that have already been researched and developed by AI scientists, from which KM specialists could benefit.

The previous chapter looked at one such area of opportunity — that is, AI's knowledge acquisition and conceptual mapping techniques vs. KM's knowledge capture and knowledge mapping approaches. This chapter examines AI's knowledge ontologies and knowledge representation approaches vs. KM's knowledge taxonomy.

Knowledge taxonomy

When developing a knowledge management system, a vocabulary of terms and relationships is often the by-product of the knowledge map. In essence, a key result of the knowledge mapping exercise is to generate a hierarchical grouping of terms that will serve as the standard vocabulary for structuring the knowledge management system for the organization.

For example, the Naval Surface Warfare Center–Carderock Division (NSWCCD), part of the U.S. Navy that does research and development in submarine and ship design, has recently undertaken a knowledge management effort in which one of the first steps was to develop a knowledge taxonomy. The knowledge taxonomy was developed by first determining the core strategic equities for the Naval Surface Warfare Center as a whole. There were seven core equities established. Based on those core equities, strategic planning key areas were identified for the Carderock Division, and then sixty key technical areas were mapped to these strategic planning key areas and core equities. Finally, about 580 knowledge areas were developed and mapped to the key technical areas. The resulting 580 knowledge areas made up the knowledge taxonomy.

Knowledge taxonomies serve as a classification of terms that are structured in some way to show relationships between these terms. In essence, they exhibit similar traits as knowledge ontologies or knowledge representation approaches that have been traditionally used in the knowledge engineering discipline. It is imperative now to examine what is meant by knowledge ontologies and show some ways of representing knowledge.

An example of an executive management knowledge taxonomy, originally developed for construction industry executives (Goodman and Chinowsky, 2000), is shown below.

Knowledge category	Focal areas
Customers and Markets	• International expansion issues and international policy
	• Market/industry characteristics
	• Stakeholder issues
	• Market analysis
	• Marketing techniques
	• Competitive trends and strategies
Leadership	• Decision making under uncertainty
	• Negotiation
	• Senior management teams
	• Pricing
	• Workplace politics
	• Working with investors
	• Ethics
	• Coaching
	• Managing organizational change
Strategy	• Strategy development techniques
	• Strategy development under uncertainty
	• Successful company strategy profiles
	• Product/process
	• Role of technology
	• Strategy execution
	• Managing changes in strategy
	• Leadership action
Managing People	• Career development
	• Workplace diversity
	• Manager performance
	• Work/family conflict
	• Female executives
	• Recruiting and training
	• Workplace regulations
Technology Management	• Technology and business strategy
	• Corporate use/impact
	• Training, learning, and resistance
	• Technology applications
	• Outsourcing

Innovation and Learning	• Types of innovation
	• Learning processes
	• Fostering learning
	• Job assignments
	• Integrating stakeholder input
	• Employee sharing
	• Rewards and incentives
	• Recruiting and training
	• Managing knowledge
	• Technology applications
	• Knowledge building
Project Management	• Redesigning and improving operations
	• Business function
	• Outsourcing
	• Supply chains
Measuring Performance	• Measurement approaches and techniques
	• Product and service performance measures
	• Non-financial performance measures
	• Finance, capital budgeting, information and analysis, risk

Knowledge ontologies

According to Gomez-Perez (1998) and Grubar (1993), an ontology is an explicit specification of a conceptualization. In the knowledge engineering community, ontologies are used to help define and structure knowledge bases and are designed for knowledge sharing and reuse purposes. There are a variety of ontologies including domain ontologies, task ontologies, and meta-ontologies. A domain ontology provides a vocabulary for describing a given domain. A task ontology defines a vocabulary for describing terms involved in problem-solving processes. A meta-ontology provides the basic terms used to codify either a domain ontology or task ontology in a formal language (Gomez-Perez, 1998).

Knowledge ontologies are used to help structure the knowledge base of an expert system similar to the way knowledge taxonomies are used to help structure the knowledge repository of a knowledge management system. To build a knowledge ontology, Gomez-Perez (1998) follows these steps:

1. Acquire knowledge.
2. Build a requirements-specification document.
3. Conceptualize the ontology.
4. Implement the ontology.
5. Evaluate during each phase.
6. Document after each phase.

A knowledge ontology or knowledge taxonomy is analogous to a blueprint for building a house. It is a schematic that will be used to structure the building. In the same way, a knowledge ontology or taxonomy will form the backbone with which to structure the knowledge management system or knowledge base.

Another type of ontology is called a knowledge representation ontology. This is derived from the knowledge engineering field and can be in terms of mainly production rules, frames, semantic networks, and cases/scripts. Production rules are in the form of IF–THEN expressions, referred to as antecedents–consequents or conditions–actions. An example is:

IF my nose is running
and I am sneezing
and my throat is sore
THEN there is a good likelihood that I have a cold

Frames are used more for declarative knowledge, as opposed to rules which are procedural in nature. Frames typically describe a collection of attributes that a given object normally possesses. For example, if a chair frame is supposed to have four legs, and a particular chair has only three, then that chair may need to be repaired (Rich, 1983).

Semantic networks describe both events and objects. They are an interconnected set of nodes with arcs depicting their relationships. For example, the Washington Redskins football team could be partially depicted in the following semantic network:

Besides semantic networks and frames, cases and scripts are also used to represent declarative knowledge. Cases and scripts are specialized types of frames that represent typical scenarios or common sequences of events. A restaurant script could include a series of events such as entering the restaurant, being seated, ordering food, eating the food, paying the bill, and exiting the restaurant. Cases are used in case-based reasoning by applying analogy to match a new situation with similar cases from previous situations.

Knowledge taxonomies in knowledge management exhibit features similar to knowledge ontologies in knowledge engineering. A knowledge taxonomy or ontology is the vocabulary and framework for structuring the knowledge management system or knowledge base. Knowledge representation approaches can be applied to portray this taxonomy or ontology. The next chapter compares the knowledge management life cycle, specifically knowledge audits, with the knowledge engineering life cycle.

References

Gomez-Perez, A. (1998), "Knowledge Sharing and Reuse," *The Handbook of Applied Expert Systems*, Ed. J. Liebowitz, CRC Press, Boca Raton, FL.

Goodman, R. and P. Chinowsky (2000), "Taxonomy of Knowledge Requirements for Construction Executives," *Journal of Management in Engineering*, January/February.

Grubar, T. (1993), *A Translation Approach to Portable Ontology Specifications, Knowledge Acquisition*, Vol. 5, Academic Press, New York.

Rich, E. (1983), *Artificial Intelligence*, McGraw-Hill, New York.

chapter four

The knowledge management life cycle vs. the knowledge engineering life cycle

In building a knowledge-centric organization, there are some basic elements needed for constructing a knowledge management pyramid. Figure 4.1 shows this pyramid.

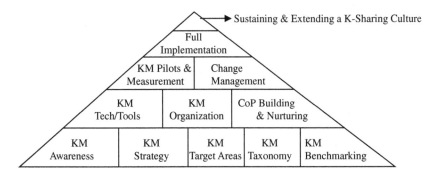

An awareness of knowledge management (KM) must first be created within the organization. This can take the form of weekly/monthly seminars, vendor visits, KM conference attendance, and other similar activities. Benchmarking what others have done in the KM field that relates to the organization's mission and competition is also very important for developing an appropriate KM strategy and program plan for the organization. As part of the KM strategy, specific target areas may be highlighted for developing KM initiatives. These targets may be ones where key expertise may be at risk of being lost due to retirements. Developing a KM taxonomy is also a critical step for structuring the knowledge management system, as has been previously discussed.

After these basic building blocks are constructed, the appropriate KM technology and tools should be selected, and the KM organizational infrastructure for directing the KM initiatives should be developed. Communities of practice (CoP) will ultimately be built and nurtured through sharing lessons learned, tips, best/worst practices, etc. within groups of people with similar interests. KM pilot projects are typically initiated with a measurement and evaluation phase to follow before moving toward full implementation. Additionally, change management focus groups should be conducted throughout the organization, because the KM initiatives may alter the way the organization performs its processes. A motivate-and-reward system and appropriate knowledge sharing proficiencies should also be incorporated to sustain and extend a knowledge sharing culture within the organization.

Knowledge management methodologies

The current literature includes several examples of knowledge management methodologies such as (Liebowitz, 2000):

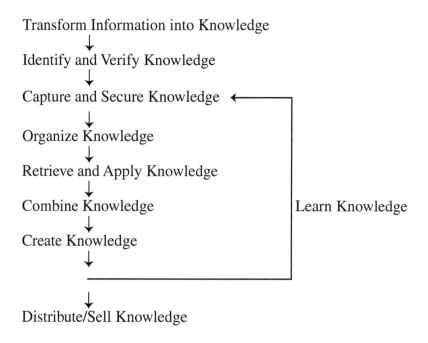

A second knowledge management methodology is suggested by Liebowitz and Beckman (1998):

Stage 1:   IDENTIFY
              Determine core competencies, sourcing strategy, and knowledge domains.

Stage 2:   CAPTURE
            Formalize existing knowledge.
Stage 3:   SELECT
            Assess knowledge relevance, value, and accuracy. Resolve
            conflicting knowledge.
Stage 4:   STORE
            Represent corporate memory in knowledge repository with
            various knowledge schema.
Stage 5:   SHARE
            Distribute knowledge automatically to users based on inter-
            est and work. Collaborate on knowledge work through vir-
            tual teams.
Stage 6:   APPLY
            Retrieve and use knowledge in making decisions, solving
            problems, automating or supporting work, job aids, and
            training.
Stage 7:   CREATE
            Discover new knowledge through research, experimenting,
            and creative thinking.
Stage 8:   SELL
            Develop and market new knowledge-based products and
            services.

Marquardt (1996) presents a simple, four-step approach:

1. Acquisition
2. Creation
3. Transfer and utilization
4. Storage

as does Wiig (1993a):

1. Creation and sourcing
2. Compilation and transformation
3. Dissemination
4. Application and value realization

The methodology suggested by Van der Spek and Spijkervet (1997) is also
a four-step approach:

1. Developing new knowledge
2. Securing new and existing knowledge
3. Distributing knowledge
4. Combining available knowledge

Ruggles' approach (1997) is slightly more complex:

1. Generation
   - Creation
   - Acquisition
   - Synthesis
   - Fusion
   - Adaptation
2. Codification
   - Capture
   - Representation
3. Transfer

O'Dell (1996) employs a straightforward seven-step method:

1. Identify
2. Collect
3. Adapt
4. Organize
5. Apply
6. Share
7. Create

The methodology proposed by Holsapple and Joshi (1997) involves only six stages, but each stage is presented in specific detail regarding the steps involved.

1. Acquiring Knowledge
   - Extracting
   - Interpreting
   - Transferring
2. Selecting Knowledge
   - Locating
   - Retrieving
   - Transferring
3. Internalizing Knowledge
   - Assessing
   - Targeting
   - Depositing
4. Using Knowledge
5. Generating Knowledge
   - Monitoring
   - Evaluating
   - Producing
   - Transferring
6. Externalizing Knowledge
   - Targeting

- Producing
- Transferring

Dataware Technologies (1998) presents its seven steps to implementing knowledge management.

1. Identify the business problem
2. Prepare for change
3. Create the KM team
4. Perform the knowledge audit and analysis
5. Define the key features of the solution
6. Implement the building blocks for KM
7. Link knowledge to people

Finally, Van der Spek and de Hoog (1998) suggest a detailed four-step approach:

1. Conceptualize
   - Make an inventory of existing knowledge
   - Analyze strong and weak points
2. Reflect
   - Decide on required improvements
   - Make plans to improve the process
3. Act
   - Secure knowledge
   - Combine knowledge
   - Distribute knowledge
   - Develop knowledge
4. Review
   - Compare old and new situation
   - Evaluate achieved results

Reviewing these numerous knowledge management methodologies reveals that there does not seem to be a comprehensive methodology that details all the steps needed to develop a knowledge management strategy and related systems. Liebowitz et al. (2000) developed the SMART methodology for knowledge management to address the need for developing a comprehensive KM methodology. The SMART methodology uses double-loop learning and looks as follows:

STRATEGIZE→ MODEL→ ACT→ REVISE→ TRANSFER

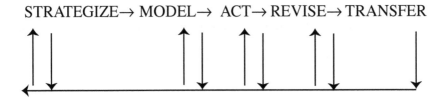

The SMART methodology and associated outputs are shown below.

Strategize:

1. Perform strategic planning.
   - Determine key knowledge requirements (i.e., core competencies).
   - Set knowledge management priorities.
2. Perform business needs analysis.
   - Identify business problem(s).
   - Establish metrics for success.
3. Conduct cultural assessment and establish a motivate and reward structure to encourage knowledge sharing.

Outputs from the strategize phase include:

- Business needs analysis document: This reviews the current IT infrastructure and documents the metrics to be used for measuring success of the knowledge management procedure.
- Cultural assessment and incentives document: This reviews the current culture of the organization and outlines approaches for encouraging knowledge sharing within the organization.

Model:

4. Perform conceptual modeling.
   - Conduct a knowledge audit.
     - Identify types and sources of knowledge (i.e., knowledge assets).
     - Determine competencies and weaknesses.
     - Perform knowledge mapping to identify the organization and flow of knowledge.
     - Perform gap analysis.
     - Provide recommendations.
   - Do knowledge planning.
     - Plan knowledge management strategy.
     - Build a supportive, knowledge sharing culture.
     - Create and define knowledge management initiatives.
     - Develop a cost-benefit analysis.
5. Perform physical modeling.
   - Develop the physical architecture.
     - Develop the framework for access, input/update, storage, and eventual distribution and use.
     - Develop a high level meta-data design.
     - Construct a visual prototype.

Outputs from the model phase include:

- Knowledge audit document: This surveys the status of knowledge in the organization and emphasizes identifying core competencies and weaknesses.
- Visual prototype: This is a knowledge map showing the taxonomy and flow of knowledge.
- Knowledge management program plan: This document specifies the initiatives and programs that will be used to meet knowledge management goals.
- Requirements specifications document: This document identifies the technological requirements for the knowledge management system (i.e., hardware and software).

Act:

6. Capture and secure knowledge.
    - Collect and verify knowledge.
    - Evaluate the knowledge.
7. Represent knowledge.
    - Formalize how the knowledge is represented.
    - Classify the knowledge.
    - Encode the knowledge.
8. Organize and store knowledge in the knowledge management system.
9. Combine knowledge.
    - Retrieve and integrate knowledge from the entire organization.
10. Create knowledge.
    - Have an open discussion with customers and interested parties both internal and external to the organization.
    - Perform exploration and discovery.
    - Conduct experimentation (i.e., trial and error).
11. Share knowledge.
    - Distribute knowledge.
    - Make knowledge easily accessible.
12. Learn knowledge and go back to Step 6.

Outputs from the act phase include:

- Knowledge acquisition document: This contains the methods and presumptions used in the process of acquiring knowledge for the knowledge management system based on the findings in the knowledge audit and the knowledge management program plan.
- Design document: This contains the knowledge classification and encoding system as well as high-level knowledge relationships. This

document also contains the design of knowledge mapping into a computer system (i.e., file structures).

- Visual and technical knowledge management system prototypes: These involve presentation of screen-mockups and the technical design of the knowledge management system.

Revise:

13. Pilot operational use of the knowledge management system.
14. Conduct knowledge review.
    - Perform quality control.
        - Review knowledge for validity and accuracy.
        - Update knowledge.
    - Perform relevance review.
        - Prune knowledge and retain what is relevant, timely, accurate, and proven useful.
15. Perform knowledge management system review.
    - Test and evaluate achieved results.
    - Revalidate/test against metrics.

Outputs from the revise phase include:

- Evaluation methodology and results document: This is for a general evaluation and review of the KM system. This document will evaluate the fitness of the developed KM system for implementation in the transfer phase. Critical analysis of the completed KM system, which includes the determination of whether the program is ready for transfer, will be completed and recommendations to continue development will be evaluated. The documentation of the evaluation methodologies used for the review and the documented results of the review are required.
- Knowledge management system prototype II: This is a preproduction, fully functional release of the KM system.
- User's guide for knowledge management system: The methods and procedures developed for the KM system are compiled into a guide for use as a training document and for the coordination of standard practices. The guide should describe both internal system processes and how the system interacts with the environment.

Transfer:

16. Publish knowledge.
17. Coordinate knowledge management activities and functions.
    - Create integrated knowledge transfer programs.
    - Document where knowledge is located and lessons learned.

- Perform serious anecdote management (i.e., publicize testimonials of the benefits of the knowledge management system).
18. Use knowledge to create value for the enterprise.
    - Sell (e.g., package knowledge bases for sale).
    - Apply (e.g., knowledge management consulting services, methodology).
    - Use (e.g., improve customer satisfaction, employee support, and training).
19. Monitor knowledge management activities via metrics.
20. Conduct post-audit.
21. Expand knowledge management initiatives.
22. Continue to learn and go back through the phases.

Outputs from the transfer phase include:

- Maintenance document for KM system: Following the completion of the final version of the KM system, documentation describing the general maintenance and change process for the system is created.
- Fully functional KM system: This refers to the final delivered and installed KM system.
- Post-audit document: Following the completed transfer of the KM system, a follow-up audit of the entire process is completed. This includes all lessons learned, user experiences, best/worst practices, and proposed changes to the methodology and/or KM system. The post audit will also include proposals for new initiatives and enhancements for the system.
- Lessons learned document: Lessons learned and other appropriate learning functions will be formatted and loaded into the appropriate corporate memory location for dissemination throughout the organization.

The knowledge audit

An important component of the SMART methodology is conducting a knowledge audit (part of the model phase). The knowledge audit looks at a targeted area and identifies which knowledge is needed and available for that area, which knowledge is missing, who has the knowledge, and how that knowledge is being used. Wiig (1993b) points out several knowledge analysis methods that can be used in a knowledge audit:

- Questionnaire-based knowledge surveys: used to obtain broad overviews of an operation's knowledge status
- Middle management target group sessions: used to identify knowledge-related conditions that warrant management attention
- Task environment analysis: used to understand, often in great detail, which knowledge is present and its role

- Verbal protocol analysis: used to identify knowledge elements, fragments, and atoms
- Basic knowledge analysis: used to identify aggregated or more detailed knowledge
- Knowledge mapping: used to develop concept maps as hierarchies or nets
- Critical knowledge function analysis: used to locate knowledge-sensitive areas
- Knowledge use and requirements analysis: used to identify how knowledge is used for business purposes and determine how situations can be improved
- Knowledge scripting and profiling: used to identify details of knowledge intensive work and the role knowledge plays in delivering quality products
- Knowledge flow analysis: used to gain an overview of knowledge exchanges, losses, or inputs of the task business processes or the whole enterprise

Certainly, a knowledge map showing the taxonomy and flow of knowledge is a critical part of the knowledge audit. Some people (e.g., Snowden, 1999) believe that the best representations for knowledge maps are stories — they convey the context, the values, and the message. The most productive audit activities may be identification of knowledge opportunities for connecting to customers, capturing the corporate memory (helping learning and preventing repeated errors), and compiling a directory of true experts and their interests.

According to Dataware (1998), one of the leaders in the knowledge management field, a productive knowledge audit need only concentrate on answering the following question: In order to solve the targeted problem, what knowledge do I have, what knowledge is missing, who needs this knowledge, and how will they use the knowledge?

The audit begins by breaking that information into two categories: what knowledge currently exists and what knowledge is missing. Once the location or source of the missing information is identified, the auditors can begin to structure the relevant information so that it can be easily found. At the conclusion of the knowledge audit, the knowledge management team has the information necessary to design its knowledge management system on paper.

Dataware strongly believes in the need to capture tacit knowledge, which is why identifying tacit knowledge is a focus of the knowledge audit. One approach is to make tacit knowledge more accessible by capturing it as metadata (data about an explicit knowledge asset). One can ask subject-matter experts what kinds of questions they most often ask others in the process of doing their jobs. One can also create skills databases, online communities of practice, and searchable repositories of resumes or skill profiles.

Consider using a qualitative organization of knowledge assets, allowing a search by topic vs. location. One can use a process-oriented approach to have a generalized model of how a business functions — from understanding customers and markets to managing people, processes, and resources — and map it to the knowledge contained in the organization. Conceptual models are often the most useful method of classification, but they are harder to construct and maintain. Conceptual models organize information around topics such as proposals, customers, or employees.

Objective of the knowledge audit (adapted from Dataware, 1998)

In order to solve the targeted business problem, the objective of a knowledge audit is to determine what knowledge we have, what knowledge is missing, who needs this knowledge, and how will the knowledge be used? The knowledge audit steps to follow are:

1. Identify what knowledge currently exists in the targeted area:
   - Determine existing and potential sinks, sources, flows, and constraints in the targeted area, including environmental factors that could influence the targeted area.
   - Identify and locate explicit and tacit knowledge in the targeted area.
   - Build a knowledge map of the taxonomy and flow of knowledge in the organization in the targeted area. The knowledge map relates topics, people, documents, ideas, and links to external resources, in respective densities, in ways that allow individuals to quickly find the knowledge they need.
2. Identify what knowledge is missing in the targeted area:
   - Perform a gap analysis to determine what knowledge is missing to achieve business goals.
   - Determine who needs the missing knowledge.
3. Provide recommendations from the knowledge audit to management regarding the status quo and possible improvements to the knowledge management activities in the targeted area.

The next section presents a knowledge audit instrument developed by Liebowitz et al. (2000) to address the steps of the knowledge audit process above. It is important to note that follow up questions via interviews, focus groups, and observation will need to be addressed after completion and analysis of this knowledge audit questionnaire.

Knowledge audit instrument

The audit instrument consists of two sets of questions. The first set seeks to find answers for Step 1 of the knowledge audit, and the second set seeks to find answers for Step 2. Step 3 of the audit would then follow from the findings in Steps 1 and 2.

Note that "categories of knowledge" refers to the types of skills, rules, and practices (i.e., things needed to be known) for doing a job. Examples might include (1) a professor who needs knowledge about presenting information, motivating students, developing course materials, ordering textbooks, etc.; (2) a lawyer who needs knowledge about research, communication skills, locating relevant documents and legislation, filing documents, court rules, ethical rules, etc.; or (3) a doctor who needs knowledge about diagnosing illnesses, performing surgery, treating diseases, bedside manner, writing prescriptions, etc. Generally speaking, information answers who, what, where, and when questions, whereas knowledge answers how and why questions.

Step 1. Identify what knowledge currently exists in a targeted area.

Questions: Answer as completely as possible

1. List specifically the categories of knowledge you need to do your job.
2. Which categories of knowledge listed in question 1 are currently available to you?

For each category of knowledge you specified in question 1:

3. How do you use this knowledge? Please list specific examples.
4. From how many sources can you obtain the knowledge? Which sources do you use? Why?
5. Besides yourself, who else might need this knowledge?
6. How often would you and others cited in question 5 use this knowledge?
7. Who are potential users of this knowledge who may not be getting the knowledge now?
8. What are the key processes that you use to obtain this knowledge?
9. How do you use this knowledge to produce a value added benefit to your organization?
10. What are the environmental/external influences impacting this knowledge?
11. What would help you identify, use, or transform this knowledge more effectively?
12. Which parts of this knowledge do you consider to be (a) in excess/abundance, (b) sparse and (c) ancient/old/past its useful life?
13. How is knowledge currently being delivered? What would be a more effective method for delivering knowledge?
14. Who are the experts in your organization housing the types of knowledge that you need?
15. In what form is the knowledge that you have gained from the experts?

16. What are the key documents and external resources that you use or would need to make your job easier?
17. What are the types of knowledge that you will need as a daily part of your job (a) in the short term (1–2 years) and (b) in the long term (3–5 years)?

Step 2. Identify what knowledge is missing in the targeted area.

Questions: Answer as completely as possible

1. What categories of knowledge do you need to do your job better?
2. What categories of knowledge do you reuse? Are there other instances where knowledge is not typically reused, but reuse would be helpful?

For each category of knowledge you specified in question 1:

3. To what degree could you improve your level of performance by having access to all of the knowledge cited in question 1?
4. Who or what might serve as potential sources of this knowledge?
5. What types of questions do you have to which you cannot find answers?

For each type of knowledge listed in question 5:

6. Of the knowledge that is missing, which types are related to: (a) job performance, (b) competitive advantage of the organization, (c) possibly leading to future expansion of the organization, or (d) simple administrative questions?
7. Which departments/people did you think would answer your question(s) but did not?
8. In what areas do you find yourself asking the same types of questions repeatedly?
9. Who have asked questions (that you are aware of) that have not been answered? In what department do they work? What level are they (i.e., job title)?
10. Which people/departments have contacted you for information?

For each person/department listed in question 10:

11. What level in the organization is each requester?
12. Is the requester a new employee (less than 1 year), a medium-term employee (1–3 years), or a long-term employee (over 3 years)?
13. Of the questions that you have been asked by others in the organization, what knowledge was requested that you consider to be (1)

essential for business performance, (2) essential for the company's competitive advantage, (3) important for leading to innovations and new business areas in the future, and (4) outdated and no longer useful for the business?

14. What mechanisms might be helpful for encouraging knowledge sharing and transfer in your organization?
15. Which aspects of your organization seem to provide barriers to effective knowledge management (i.e., what constraints impede knowledge sharing and transfer)?
16. What are the main reasons that you could have made errors/mistakes on the job?

If your organization has considered outsourcing in the last 5 years:

17. In what areas was the outsourcing considered?
18. If outsourcing was rejected, why?
19. If outsourcing has taken place, why?
20. How much time do you spend looking for knowledge?

Appendix B shows part of the results from a knowledge audit conducted by Liebowitz, Rubenstein-Montano, and Buchwalter at the Social Security Administration.

Knowledge engineering life cycle

The knowledge management life cycle and related methodologies overlap with the knowledge engineering life cycle used to build knowledge bases and knowledge-based systems. The knowledge engineering life cycle typically has the following phases:

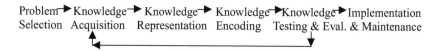

The problem selection phase looks at scoping the problem to select a well-defined, well-bounded area of knowledge for developing the knowledge-based system. This is similar to identifying the KM target areas as shown previously in the knowledge management pyramid. Once the area of knowledge is selected, knowledge acquisition is used to elicit the knowledge from the expert(s) and from other sources (e.g., documents, manuals, technical reports). Interviewing techniques, verbal walkthroughs (protocol analyses), observations, and other knowledge acquisition techniques are used for eliciting knowledge from the expert(s). After acquiring the knowledge, it is then represented as procedural, declarative, case-based, or other types of knowledge (as discussed in the previous chapter). The represented knowledge is then encoded, tested, and evaluated. Knowledge refinements will need to be iteratively made and then new knowledge will need to be acquired, represented, encoded, and tested. This is an iterative, rapid prototyping approach which is typically used in knowledge engineering and systems development. Eventually, the knowledge-based system is fully developed and implemented within the organization.

Another popular knowledge engineering methodology is CommonKADS (Schreiber et al., 2000). According to Schreiber and associates, knowledge engineering offers many useful concepts and methods for knowledge management:

- Knowledge-oriented organization analysis helps to quickly map out fruitful areas for knowledge management actions.
- Task and agent analysis has proven useful for clarifying knowledge bottlenecks in specific areas.
- Knowledge engineering places strong emphasis on the conceptual modeling of knowledge intensive activities.
- The experience of knowledge engineering shows that there are many recurring structures and mechanisms in knowledge work.
- CommonKADS has been used in knowledge management quick scans and workshops as well as in IT strategy scoping and feasibility projects, and gives a sound support in the early stages of requirements elicitation and specification in systems projects.

There are several parallels between the knowledge engineering life cycle and the knowledge management life cycle. The problem selection and scoping phase in knowledge engineering is analogous to parts of the strategize and model phases in the SMART knowledge management methodology. Targeted areas need to be identified in both knowledge management and knowledge engineering applications. The knowledge acquisition and representation phases in the knowledge engineering cycle are similar to the knowledge capture (especially the capture of tacit knowledge) and development of the knowledge management system under the act phase of the SMART methodology. The knowledge engineering testing and evaluation phase has commonalities with the revise phase in SMART, and the implementation phase in knowledge engineering is similar to the transfer phase in the SMART methodology. Thus, knowledge engineering has direct ties to knowledge management, and the techniques and tools used in knowledge engineering can help advance the knowledge management field.

References

Dataware Technologies (1998), Seven Steps to Implementing Knowledge Management in Your Organization, Corporate Executive Briefing, www.dataware.com.

Holsapple, C. and K. Joshi (1997), "Knowledge Management: A Three-Fold Framework," Kentucky Initiative for Knowledge Management, Paper 104.

Liebowitz, J. (2000), *Building Organizational Intelligence: A Knowlege Management Primer*, CRC Press, Boca Raton, FL.

Liebowitz, J., B. Rubenstein-Montano, J. Buchwalter, et al. (2000), "The Knowledge Audit," *Journal of Knowledge and Process Management*, Vol. 7, No. 1.

Liebowitz, J. and T. Beckman (1998), *Knowledge Organizations: What Every Manager Should Know*, CRC Press, Boca Raton, FL.

Marquardt, M. (1996), *Building the Learning Organization*, McGraw-Hill, New York.

O'Dell, C. (1996), "A Correct Review of Knowledge Management Best Practice," Conference on Knowledge Management and the Transfer of Best Practices, Business Intelligence, London, UK.

Ruggles, R. (1997), "Tools for Knowledge Management: An Introduction," in *Knowledge Management Tools*, Butterworth-Heinemann, Cambridge, MA.

Schreiber, G., H. Akkermans, A. Anjewierden, R. de Hoog, N. Shadbolt, W. van de Velde, and B. Wielinga (2000), *Knowledge Engineering and Management: The CommonKADS Methodology*, MIT Press, Cambridge, MA.

Snowden, D. (1999), "Story Telling for Knowledge Capture," The International Knowledge Management Summit Proceedings, The Delphi Group, San Diego, CA, March 29-31.

Van der Spek, R. and R. de Hoog (1998), *Knowledge Management Network*, University of Amsterdam Press, Amsterdam.

Van der Spek, R. and A. Spijkervet (1997), "Knowledge Management: Dealing Intelligently with Knowledge," in *Knowledge Management and Its Integrative Elements*, Eds. J. Liebowitz and L. Wilcox, CRC Press, Boca Raton, FL.

Wiig, K. (1993a), *Knowledge Management Foundations*, Schema Press, Arlington, TX.

Wiig, K. (1993b), *Knowledge Management Methods*, Schema Press, Arlington, TX.

chapter five

Knowledge-based systems and knowledge management

A key component of a knowledge management strategy is typically a Web-based expert and knowledge retention system. This part of the knowledge management system captures the thought making processes of key experts before they retire. This is essentially the purpose of a knowledge-based or expert system. Expert systems thus play a key role in knowledge management. Alternatively, there is software, like Virage's system, that allows online searchable video. In this manner, an expert can be interviewed and videotaped, and his/her knowledge base (per the video) can be searched. CNN uses this type of capability, and Sandia National Labs uses a similar feature in their knowledge preservation project.

The knowledge engineering steps (as discussed in an earlier chapter) for building a knowledge-based/expert system are fairly universal, as shown in DiStasio, Droitsch, and Medsker's (1999) expert system development life cycle:

1. Problem identification and description
2. Domain understanding and knowledge acquisition
3. Structure design and knowledge organization
4. System development
5. Verification and validation
6. Operations and maintenance

There are typically four major components of an expert system. The dialog structure serves as the user interface between the user and the expert system to allow the user to run through a sample session. The control structure of the expert system is called the inference engine. This problem-processing component allows for the generation and testing of various hypotheses, or goals, to test their truth. Search strategies are contained within the inference engine in order to derive and test various goals. The knowledge base is the third major part of the expert system and contains the facts and rules of thumb in a given area. The inference engine uses the knowledge base in order to generate and test the various hypotheses. The last component of the expert system is the explanation facility which allows the user to ask how and why questions to see how the expert system arrives at its conclusions.

To obtain a better understanding of how a Web-based expert system is built, Rubenstein-Montano et al. (2000) developed such a system for the Health Care Financing Administration (HCFA) to identify information technology (IT) system impact areas for proposed new projects at HCFA. This Web-based expert system (http://rocket.ifsm.umbc.edu/hcfa) captured the critical knowledge of seven to ten experts associated with the seven IT system impact areas (e.g., security, data center, telecom) that could relate to a new project at HCFA.

The HCFA IT System Impact Advisor (HITSIA) case study

The knowledge engineering life cycle of HITSIA followed the steps identified in the previous chapter, namely, problem selection, knowledge acquisition, knowledge representation, knowledge encoding, and knowledge testing and evaluation.

The IT personnel at HCFA were being inundated with calls and meetings with project owners in terms of determining the effect of the proposed projects in the IT areas. To help alleviate this burden, HITSIA was developed so that project owners could interact directly with the system. The expert system would act as a surrogate expert to help determine whether security, HCFA data center, LAN/WAN (local area networks/wide area networks), Web, data and database administration, telecommunications, and IT architecture impacted their proposed project. The objective of HITSIA was to develop a Web-based expert system to identify IT system impact areas (and level of impact — high, medium, or low) for new HCFA projects.

After performing problem selection and scoping, the next step was knowledge acquisition. The HITSIA knowledge engineers conducted numerous structured interviews with the HCFA sponsors and experts. In addition to those knowledge acquisition sessions, various pertinent HCFA documents were reviewed and many follow-up discussions via e-mail ensued. This process was iteratively performed as the knowledge base was built.

Once knowledge was acquired in the various IT system impact areas from the knowledge acquisition sessions with the respective experts, the next step involved representing the knowledge for encoding purposes. A rule-based format seemed to be the most natural way of representing the domain knowledge, primarily because the experts couched their thoughts in terms of IF–THEN clauses. The taped knowledge acquisition sessions and corresponding notes were converted into decision trees and then into a set of rules for each IT system impact area.

After representing the knowledge as rules, the next step involved knowledge encoding. This stage dealt with encoding the rules in the knowledge base using a commercially available expert system shell (one of the requirements of the effort). We decided to use Multilogic's Exsys Developer/Resolver for developing the expert system and Multilogic's Exsys Web Runtime/Netrunner for converting the expert system into a Web-based version. The U.S. Department of Labor has used these tools for their ten or more Web-based expert systems with great success (www.dol.gov/elaws). For integration purposes into HCFA's IT investment database, an ASP (active server pages) wrapper was developed to take the recommendations from HITSIA and encode them within HCFA's database. Figure 5.1 shows sample conclusions from HITSIA based upon around 25 questions.

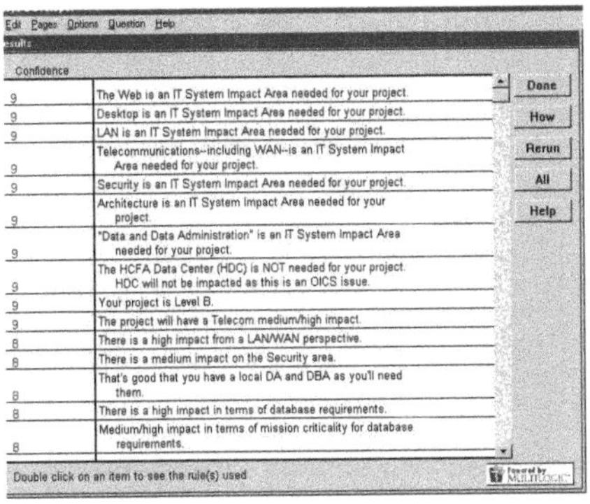

Figure 5.1   Sample HITSIA conclusions.

The testing and evaluation of HITSIA have been conducted in numerous ways. First, we applied the validation function within the Resolver/Netrunner shell to test for uncovered paths, inconsistency, and unused qualifiers and rules. We showed initial versions of HITSIA to our sponsor and her colleagues at HCFA. We also had the experts use HITSIA for several weeks to gauge their responses and incorporate their comments into the final version of HITSIA. We gave demos of HITSIA to the sponsor and experts over several hours so that we could have a direct exchange of ideas. By placing HITSIA on the Web server, the experts could use HITSIA and run sample sessions at their leisure.

The drug and alcohol abuse advisor

Liebowitz developed an expert system prototype, using Exsys/Resolver, for helping UMBC (University of Maryland — Baltimore County) students become better familiar with the legal consequences of and awareness creation relating to drug and alcohol abuse. Figure 5.2 shows a sample user session with this expert system.

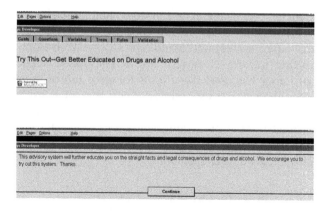

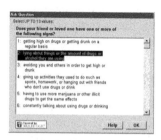

Figure 5.2   Sample user run with the Drug and Alcohol Abuse Advisor.

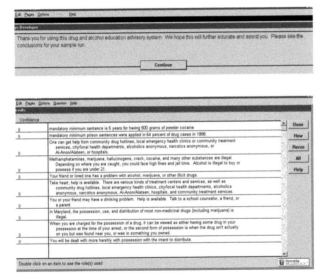

Figure 5.2   (Continued)

Knowledge-based or expert systems can be used to capture the knowledge and experiential learning of human experts in well-defined areas. Unfortunately, most of these expert systems are fairly static and do not usually have learning capabilities. The use of intelligent agents can be part of the knowledge management spectrum for improved knowledge coordination, collaboration, and dissemination (and, perhaps, knowledge creation). The next chapter will look at the use of intelligent and multi-agents to serve this purpose.

References

Rubenstein-Montano, B., J. Liebowitz, M. Gross, M. Imhoff, T. Richburg, and M. Brown (2000), HIPP: HCFA IT Project Planner, *Journal of Computer Information Systems*, Vol. 40, No. 4.

DiStasio, M., R. Droitsch, and L. Medsker (1999), "Web-Based Expert Systems for Elaws," *Failure and Lessons Learned in Information Technology Management: An International Journal*, Vol. 3, No. 2.

chapter six

Intelligent agents and knowledge dissemination

Multi-agent system frameworks

Although many intelligent agent-based applications have been successfully developed and deployed in real world settings, most of them are single intelligent agent-based systems instead of multi-agent systems. A single intelligent agent system, when a new task is delegated by the user, determines the goal of the task, evaluates how the goal can be reached in an effective manner, and performs the necessary actions by itself. In addition, a single intelligent agent system is capable of learning from past experience and responding to unforeseen situations with its reasoning strategies. It is reactive/autonomous so that it can sense the current state of its environment and act independently to make progress toward its goals.

While most intelligent agents provide their users with significant value when used in isolation, there is an increasing demand for programs that can collaborate — to exchange information and services with other agents, thereby increasing knowledge cooperation and dissemination — which is a critical function of knowledge management. Intelligent agent technology can help direct ("push") lessons learned to appropriate individuals in the organization versus having to use a passive, "pull" knowledge dissemination approach whereby users of the knowledge management system have to search for relevant lessons in their area of knowledge.

Over the past few years, some interesting work has been developed for creating multi-agent system frameworks. One such framework by DeLoach (1999) develops a methodology for multi-agent systems engineering. The framework includes the following:

1. Identify agent types.
2. Identify the possible interactions between agent types.
3. Define coordination protocols for each type of interaction.
4. Map actions identified in agent conversations to internal components.

5. Define inputs, flows, and outputs associated with the agents.
6. Select the agent types that are needed.
7. Determine the number of agents required of each type and define the agents' physical locations or addresses, the types of conversations that agents will be able to hold, and any other parameters defined in the domain.

Zeus, developed at British Telecom Laboratories by Collis et al. (1998), is an advanced toolkit for engineering distributed multi-agent systems. Zeus contains an agent component library, visualization tools, and agent building software. The Zeus agent design methodology is to determine candidate agents, define each agent using the graphical Zeus Generator tool, and identify tasks, describe agent relationships using Zeus Generator, choose from a list of prewritten coordination strategies, and implement/encode the agents. OAA (Open Agent Architecture) by SRI and JADE (Java Agent Development Environment) are two other multi-agent toolkits that have been actively used.

Flores-Mendez (1999), with the Collaborative Agents Group at the University of Calgary, proposes the need for a standardized multi-agent system framework. He describes the multi-agent system as an environment consisting of areas. Each area is required to have exactly one local area coordinator, which is an agent that acts as a facilitator for other agents within its area. Agents use the services of local area coordinators to access other agents in the system. Agents can also be connected with yellow page servers and cooperation domain server agents.

A variety of other work on multi-agent systems has been performed as well. Landauer and Bellman (1999) describe an approach to integration involving constructing complex systems that rely on cooperative collections of agents instead of a central planner or organizer. Sycara and Zeng (1996) discuss the coordination of multiple intelligent software agents. Arisha et al. (1999), from the University of Maryland–College Park, describe a platform called Impact for collaborating agents. Labrou, Finin, and Peng (1999), at the University of Maryland–Baltimore County (UMBC), describe the various agent communications languages — KQML (knowledge query manipulation language), FIPA ACL (Foundation for Intelligent Physical Agents — agent communication language), and others. Joshi and Singh (1999) guest edited a special issue of *Communications of the ACM* which dealt with multi-agent systems on the Internet and included a myriad of papers looking at multi-agent system frameworks and applications.

Furthermore, Sycara (1998) discusses multi-agent systems and the challenges ahead, namely: (1) how to decompose problems and allocate tasks to individual agents, (2) how to coordinate agent control and communications, (3) how to make multiple agents act in a coherent manner, (4) how to make individual agents reason about other agents and the state of coordination, (5) how to reconcile conflicting goals between coordinating agents, and (6) how to engineer practical multi-agent systems. In addition to this list of challenges, many researchers are looking at only autonomous agents, but in many situations, the integration of human collaboration with agent-based interaction will be crucial. Researchers such as Volksen et al. (1996) have developed Cooperation-Ware as a framework for human-agent collaboration.

Multi-agent COTR system (MACS)

MACS is a multi-agent system being developed to help the contracting officer's technical representative (COTR — the government representative) answer questions relating to the pre-award phase of a contract in the area of defense procurement and contracting (Liebowitz et al., 2000). Specifically, MACS tries to answer questions relating to what forms are needed in a procurement package (forms agent), what type of synopsis is required (synopsis agent), how to evaluate proposals (evaluation agent), what type of contract is recommended (contracts agent), and whether to go sole source, i.e., with just one vendor (justification and approval agent).

Agent communication and interaction in MACS proceeds in the following manner:

1. The user agent welcomes the user.
2. User sends a user request to the user agent (currently via predetermined keywords selected from a list).
3. The user agent determines if it understands the request, and if so, it then broadcasts the request to the specialty agents.
4. If the user agent needs further clarification from the user, it sends the request back to the user for further clarification.
5. The user then sends the clarified request to the user agent, who, in turn, sends it to the specialty agents.
6. If a specialty agent can answer the request, it sends the answer back to the user agent, who, in turn, forwards it to the user.
7. If a specialty agent cannot answer the request, it sends the request back to the user agent, who then (if appropriate) forwards it to the user for further clarification.
8. If, after several rounds of clarification, none of the specialty agents can determine an answer to the request, this information is sent to the user agent, who, in turn, sends this reply to the user.

Figure 6.1 illustrates the system architecture and communication described above. Figures 6.2 and 6.3 show screen shots of the input and output screens in MACS.

Currently, MACS has been ported and encoded from Reticular Systems' AgentBuilder to SRI's OAA (Open Agent Architecture). A natural language processing capability for the user agent has also been encoded, as well as a Bayesian learning algorithm so that the user agent can better direct a user query to the most appropriate specialty agent.

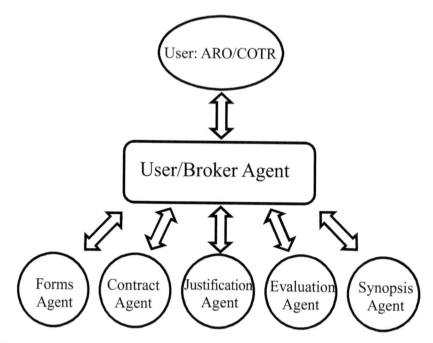

Figure 6.1 Agent architecture and communication.

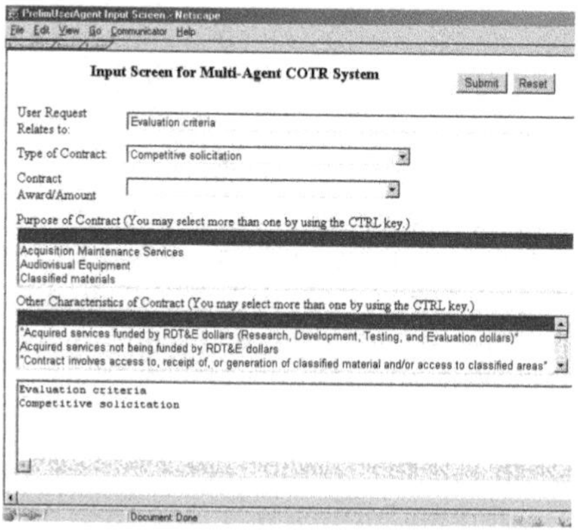

Figure 6.2 Input screen for MACS.

A number of researchers, such as Aguirre et al. (2001), have referred to the integration of multi-agents and knowledge management as multi-agent based knowledge networks. In dealing with knowledge sharing and knowledge filtering/searching, multi-agent technology can greatly enhance the current state of the art in knowledge management. Instead of having a passive collection, analysis, and dissemination of lessons learned, for example, in a lessons learned knowledge repository, intelligent agent technology

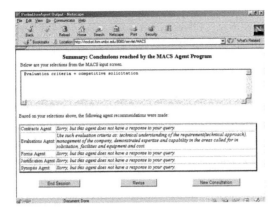

Figure 6.3   MACS output screen.

can help actively collect, analyze, and share the appropriate lessons with individuals in the organization (or externally, if warranted), in the near future, intelligent/multi-agent technology will probably be the greatest AI contribution to the knowledge management field.

References

Aguirre, J., R. Brena, and F. Cantu (2001), "Multi-Agent Based Knowledge Networks," *Expert Systems with Applications: An International Journal*, Vol. 20, No. 1.

Arisha, K., F. Ozcan, R. Ross, V. Subrahmanian, T. Eiter, and S. Kraus (1999), "Impact: A Platform for Collaborating Agents," *IEEE Intelligent Systems*, IEEE Computer Society Press, Los Alamitos, CA, March/April.

Collis, J., H. Nwana, D. Ndumu, and L. Lee (1998), "Zeus: An Advanced Toolkit for Engineering Distributed Multi-Agent Systems," British Telecom Laboratories, Marlesham Heath, UK, www.labs.bt.com/projects/agents/.

DeLoach, S. (1999), "Multiagent Systems Engineering: A Methodology and Language for Designing Agent Systems," Proceedings of Agent-Oriented Information Systems.

Flores-Mendez, R. (1999), "Towards a Standardized Multi-Agent System Framework," *ACM Crossroads*, Summer.

Joshi, A. and M. Singh (1999), "Multiagents Systems on the Net," *Communications of the ACM*, special issue, March.

Labrou, Y., T. Finin, and Y. Peng (1999), "Agent Communication Languages: The Current Landscape," *IEEE Intelligent Systems*, March/April.

Landauer, C. and K. Bellman (1999), "Agent-Based Information Infrastructure," The Aerospace Corporation, Los Angeles, CA, April.

Liebowitz, J., M. Adya, B. Rubenstein-Montano, V. Yoon, J. Buchwalter, M. Imhoff, S. Baek, and C. Suen (2000), "MACS: Multi-Agent COTR System for Defense Contracting," *Journal of Knowledge-Based Systems*, December.

Sycara, K. (1998), "Multiagent Systems," *AI Magazine*, Summer.

Sycara, K. and D. Zeng (1996), "Coordination of Multiple Intelligent Software Agents," *International Journal of Cooperative Information Systems*, Vol. 19.

Volksen, G., H. Haugeneder, A. Jarczyk, and P. Loffler (1996), "Cooperation-Ware: Integration of Human Collaboration with Agent-Based Interaction," Siemens Corporation, Munich, Germany.

chapter seven

Knowledge discovery and knowledge management

Knowledge discovery, a subfield of artificial intelligence, can be used in the knowledge creation function in knowledge management. Knowledge discovery applies AI techniques to inductively determine new relationships, trends, and knowledge for a given application. Case-based reasoning, data/text mining, and intelligent/autonomous agents are typical techniques in knowledge discovery. This chapter discusses some projects and usage of these techniques to assist in the knowledge management function (Liebowitz, 2000, 1999; Liebowitz and Beckman, 1998).

An intelligent agent assisted performance support system for the criminal investigative analysis domain

The purpose of this research project, funded by a Maryland Industrial Partnership System (MIPS) grant, was to build and deliver a client-based, Windows environment software application that acts as an intelligent agent to assist Wisdom Builder users in the criminal investigative analysis domain. Wisdom Builder (2000) is a knowledge management tool that has been used in the intelligence and investigative domains to help users in the requirements, collection, analysis, and reporting phases. Wisdom Builder supports activities across these four major phases of the analytical research process to help monitor areas of interest and develop strategies that promote innovation, productivity, and profitability. One of the main limitations of Wisdom Builder is that it may be somewhat difficult to use, partly due to its powerful features. In order to reduce the burden on the user and help guide the user through a session of Wisdom Builder, it would be helpful to have an intelligent user agent to provide recommendations to the user on setting up a Wisdom Builder application and performing his/her requirements and analysis functions using the program. In a sense, this intelligent user agent may work like a Microsoft wizard by looking over the shoulder of the user and providing suggestions on how best to gather requirements and perform the analysis steps. The agent would reside in the background, monitoring the user's actions, and offer suggestions and/or courses of action to the user based on the user's interaction with Wisdom Builder. At the user's request, the agent could interact with Wisdom Builder directly to perform the suggested actions. This research touches on the ideas of personal agents (e.g., Gams and Hribovsek, 1996; Soltysiak and Crabtree, 1998), software coaches, intelligent user interfaces (e.g., Intelligent User Interfaces Conference Proceedings, 1999), and case-based reasoning (e.g., Munoz-Avila, Hendler, and Aha, 1999; O'Leary and Selfridge, 1999).

The application domains selected for testing purposes are intelligence analysis, competitive intelligence, and criminal investigation analysis. In Heuer's book, *Psychology of Intelligence Analysis* (1999), the analyst typically applies six key steps in the analytical process: defining the problem, generating hypotheses, collecting information, evaluating hypotheses, selecting the most likely hypothesis, and the ongoing monitoring of new information. The FBI Academy's special agent's training includes 16 weeks of intensive instruction in firearms, practical applications, physical training/defensive tactics, law, forensic science, interviewing, informant development, communications, white-collar crime, drug investigations, ethics, organized crime, behavioral science, computer skills, and national security matters. A number of artificial intelligence-based systems such as COPLINK (http://ai.bpa.arizona.edu/coplink), whose knowledge-based databases are used by the Tucson police department to provide large-scale intelligence analysis capabilities including the identification of previously unknown relationships, have been built to assist in the law enforcement area. However, a strong need exists to develop intelligent agent-assisted performance support systems to aid the investigative analyst in performing his/her critical functions.

To help elicit the analyst's requirements for an intelligent agent for the Wisdom Builder tool, a Web-based survey (encoded via Perseus Development's Survey Solution for the Web) was used to elicit the following major information:

- Name the top three features of Wisdom Builder that you find most helpful to you in your analysis.
- Name three possible ways to maximize the usability of these features.
- Name some features that you would like to have in Wisdom Builder that are not currently available to you.

The results of this survey revealed that the main features users wanted to include in the intelligent user agent were: having the agent look over the analyst's shoulder to make sure that he/she does not omit useful facts and hypotheses in solving a case, making sure that the hypotheses and resulting conclusion generated seem consistent and reasonable, helping the analyst proceed through the analysis phase and key strokes in the Wisdom Builder product, having the agent help the analyst in the thinking and reasoning processes involved in making intelligence judgments, and generally helping users respond to their questions when setting up their Wisdom Builder application.

Intelligent user agent design and development and case-based reasoning

After collecting these user requirements, the next step of the research project was to design the intelligent user agent. At first glance, it appeared that the wizard technology would be very suitable for the agent. For example, Microsoft Access includes various wizards such as the OutputWriter Wizard, ChartExcel Wizard, ReportExcel Wizard, NotesTable Wizard, Help Wizard, ReportRun Wizard, PrintExcel Wizard, CodeBox Wizard, and Renaming Wizard. Atkins' (2000) article on generating a Microsoft Wizard interface discusses categorizing a wizard into two groups: layout and functional. The main layout features of a wizard are action buttons at the bottom of the window, a graphic in an area on the left, a data area to the right of the graphic and above the buttons, and instructions displayed somewhere on the window. In addition to these layout features, a wizard interface often has the following functional features (Atkins, 2000):

- Buttons are used to move through a set of input pages.
- Instructions change for every page.
- Each page is validated before continuing to the next page.
- Each subsequent page uses the previous page's data.
- Nothing is formally committed until the whole process is complete, and the user presses the "finish" button.
- The user can press the "cancel" button at any time and leave the wizard without leaving behind partial or invalid data.
- The action buttons are activated/deactivated or the label changes based on the page context.

In many of the Microsoft wizards, case-based reasoning (CBR) is employed. CBR involves analogical reasoning whereby a new situation is compared with existing cases in a case base, matching similar case features or adapting from those cases for resolving the new situation. Bayesian user modeling has also been applied and integrated with CBR systems to infer a user's needs by considering a user's background, actions, and queries. For example, the Lumiere Project by Microsoft Research employed Bayesian user modeling and served as the basis for the Office Assistant in Microsoft Office's suite of productivity applications (Horvitz et al., 1997). Aha and Chang (1996) discuss a system at the Navy Center for Applied Research in Artificial Intelligence at the Naval Research Laboratory called INBANCA (integrating Bayes networks with case-based reasoning for planning).

The researchers selected case-based reasoning as the intelligent system methodology for the proposed agent. After trying Esteem's/SHAI's CBR Express and TecInno's (Germany) CBR-Works case-based reasoning tools, CBR-Works was chosen for the following reasons: (1) an executable/client version as well as a server version of the case-based application could be created, (2) the user interface handled some natural language processing, (3) the tool had been used by a number of major companies, and (4) the tool was fairly easy to use. The first part of this CBR application focused on answering queries arising as users set up their applications.

The next step involved creating a knowledge taxonomy that related to the main functions and tabs in Wisdom Builder (e.g., table tabs, connections tabs, timeline tabs, link notebooks tabs, create tabs, deliver tabs). Then, a listing of the typical user questions as related to each function/tab was created, and about 95 cases were then input into the case base, which would respond with the answers associated with these questions. The server version allows Wisdom Builder users to send additional questions and possible answers to WisdomBuilder, Inc., where the information is validated before confirmation of its acceptance into the case base. The server version would also promote an online community of Wisdom Builder users worldwide.

The next step was to further develop this agent by enabling it to monitor the user's actions and provide suggestions during the analysis phase. In order to provide this level of intelligence, a knowledge ontology and general model of analysis had to be developed. This would serve as the framework or model-based paradigm for performing an analysis step. The following general model was used, based upon Quade and Boucher's (1978) work:

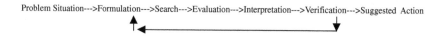

Problem Situation--->Formulation--->Search--->Evaluation--->Interpretation--->Verification--->Suggested Action

where:

Formulation	=	conceptual phase (clarifying objectives, defining issues of concern, and limiting the problem)
Search	=	research phase (looking for data and relationships as well as alternatives)
Evaluation	=	analytic phase (building various models, using them to predict the consequences that are likely to follow from each choice of alternative, and comparing alternatives)
Interpretation	=	judgmental phase (comparing the alternatives further, deriving conclusions about them, and indicating a course of action)
Verification	=	scientific phase (testing the conclusions by experimentation)

The resulting tool developed from this effort is called POINT (problem organization intelligence tool). It is encoded in Visual Basic 6.0 and MS-Access. POINT (Phillips, Liebowitz, and Kisiel, 2001) helps analysts (e.g., intelligence analysts, financial analysts, competitive intelligence analysts), structure and define their problems, identify the knowledge base, and perform analysis in determining likely hypotheses for solving the problem. Appendix C describes in depth the intelligence analysis model used in POINT.

Knowledge management and case-based reasoning

Case-based reasoning (CBR) is a good solution for the technology needs of some knowledge management (KM) tasks such as customer relationship management (Aha, 1999). Many people feel that CBR can assist in capturing tacit knowledge and can contribute to process-centered activities in support of KM. Various interactive (e.g., conversational) CBR techniques have been applied to KM applications such as customer service, aircraft maintenance, automotive part design, and planning ongoing medical practices (Aha, 1999). Most early work on mixed-initiative CBR supported interactive diagnosis focused on help desk and call center applications. In the context of KM, future focus may be on using CBR for decision support. For example, future work on embedded CBR systems may involve the user in determining what personal biases to employ in his/her KM activities (Aha, 1999). Currently, there is increasing emphasis on working with cases to capture and manage procedural knowledge. More work toward the design of embedded and modular CBR systems in which sub-processes can be retrieved, adapted, and composed to support specific KM activities will probably be performed in the near future (Aha, 1999).

Knowledge management and text mining

Text mining involves extracting patterns, behaviors, and general knowledge from large collections of textual information, which are often found in knowledge repositories as part of KM systems (Cox, 2000). The text knowledge mining process typically applies knowledge discovery techniques to develop self-organizing maps (SOMs), clusters, and predictive models (rules). An SOM brings together related concepts and shows their intensity or frequency within the data/text base as well as their proximity to other concepts. Fuzzy clusters provide a spatial analysis of documents and semantic concepts in the form of related aggregations. Fuzzy rules encapsulate the extracted knowledge to evaluate and classify new documents and make predictions based on document contents. Text mining covers deeply buried patterns in the textual data. IBM's Intelligent Miner for Text is an information extraction/text mining software package to help perform text mining functions.

Knowledge management and neural networks

Neural networks are artificial interconnected webs of neurons that try to emulate the workings of the human brain. Typically, neural networks are trained to achieve a desired level of performance. Neural networks could loosely fall into the knowledge discovery area to try to infer patterns from data, knowledge, and images. An example of a project that used a neural network approach as part of a generic intelligent scheduling toolkit (called GUESS) is briefly discussed below.

Generically used expert scheduling system (GUESS)

NASA Goddard Space Flight Center was interested in developing a generic intelligent scheduling toolkit that would be used primarily to schedule experimenter requests to use NASA Goddard-supported spacecraft. GUESS was developed by Liebowitz et al. (2000) for this application, as well as for generic scheduling purposes such as Army battalion training exercises, the arrival of military units in a deployed theatre, university courses, and base-ball games for a local municipality. GUESS has four techniques for scheduling: a suggestion tabulator based on heuristics used by human schedulers, a hill climbing algorithm, a genetic algorithm, and a neural network algorithm based on the Hopfield network.

In terms of the neural network approach, a modified Hopfield network was used to arrive at an optimal schedule given a set of events and the constraints that must be satisfied between events (see Figure 7.1).

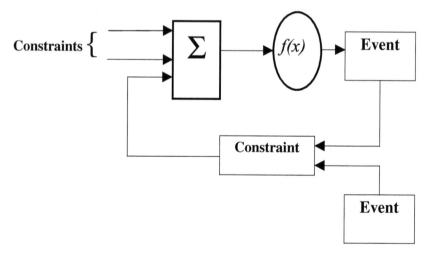

Figure 7.1   Neural network connections.

Each of the constraints on an event produces an error signal. The error signal is chosen to cause the event to move in the correct direction to produce a *satisfied* schedule. The errors on a given event induced by the constraints are summed together and then passed through a sigmoid function. The output of the sigmoid function f(x) is used to shift the begin and end times of the event to drive the schedule to a more satisfied state. Several different sigmoid functions were tested. The most promising was f(x) = tanh (x). This yielded the following equation for the neural network:

$$\delta i = k \cdot \tanh\left(\sum_j c(ei,ej)\right)$$

where,

δi = delta time to shift the ith event
k = weighting constant
$c(ei, ej)$ = constraint error between the ith event and the jth event

The neural network approach worked very well (for details, consult Liebowitz et al., 2000), as compared with the suggestion tabulator, hill climbing, and genetic algorithm approach. Generally speaking, the hill climbing algorithm performed the best (about a 10% performance improvement over the neural network approach in terms of overall schedule satisfaction), but the neural network approach took about one-third of the time to arrive at the final schedule vs. the hill climbing algorithm.

Figure 7.2 shows a sample screen shot from GUESS. As visible in the screen shot, GUESS displays a Gantt chart of scheduled events on top and a resource consumption profile down below. GUESS has been tested to schedule 10,000 events and over 36,000 constraints in less than three minutes on a Pentium I IBM laptop computer.

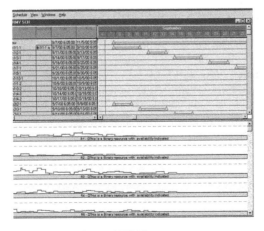

Figure 7.2 Sample screen shot from GUESS.

References

Aha, D. (1999), "The AAAI-99 KM/CBR Workshop: Summary of Contributions," Practical CBR Strategies for Building and Maintaining Corporate Memories Workshop, International Conference on Case-Based Reasoning.

Aha, D. and L.W. Chang (1996), "Cooperative Bayesian and Case-Based Reasoning for Solving Multi-Agent Planning Tasks," Technical Report, Navy Center for Applied Research in Artificial Intelligence, Naval Research Laboratory, Washington, D.C.

Atkins, K. (2000), "I'm Off to See the Wizard," ARIS Corporation, www.arrowsent.com/oratip/genwiz.htm.

Cox, E. (2000), "Free-form Text Data Mining Integrating Fuzzy Systems, Self-Organizing Neural Nets, and Rule-Based Knowledge Bases," *PC AI*, September/October.

Gams, M. and B. Hribovsek (1996), "Intelligent Personal Agent Interface for Operating Systems," *Applied Artificial Intelligence Journal*, Vol. 10, No. 4.

Heuer, R. (1999), *Psychology of Intelligence Analysis*, Center for the Study of Intelligence, Central Intelligence Agency, www.odci.gov/csi/books/19104/index.html.

Horvitz, E., J. Breese, D. Heckerman, D. Hovel, and K. Rommelse (1997), "The Lumiere Project: Bayesian User Modeling for Inferring the Goals and Needs of Software Users," Microsoft Research, Redmond, WA.

Intelligent User Interfaces Conference Proceedings (1999), Association for Computing Machinery, New Orleans.

Liebowitz, J., Ed. (1999), *The Knowledge Management Handbook*, CRC Press, Boca Raton, FL.

Liebowitz, J. (2000), *Building Organizational Intelligence: A Knowledge Management Primer*, CRC Press, Boca Raton, FL.

Liebowitz, J. And T. Beckman (1998), *Knowledge Organizations: What Every Manager Should Know*, CRC Press, Boca Raton, FL.

Liebowitz, J., I. Rodens, J. Zeide, and C. Suen (2000), "Developing a Neural Network Approach for Intelligent Scheduling in GUESS," *Expert Systems Journal*, September.

Munoz-Avila, H., J. Hendler, and D. Aha (1999), "Conversational Case-Based Planning," *Review of Applied Expert Systems*, Vol. 5.

O'Leary, D. and P. Selfridge (1999), "Knowledge Management for Best Practices," *SIGART Intelligence*, Winter.

Phillips, J., J. Liebowitz, and K. Kisiel (2001), "Modeling the Intelligence Analysis Process for Intelligent User Agent Development," *Research and Practice in Human Resources Management Journal*, National University of Singapore, January.

Quade, E.S. and W.I. Boucher (1978), *Systems Analysis and Policy Planning*, Elsevier/North-Holland, New York.

Soltysiak, S. and I. Crabtree (1998), "Automatic Learning of User Profiles — Towards the Personalization of Agent Services," *British Telecom Technology Journal*, Vol. 16, No. 3.

chapter eight

People and culture: lessons learned from AI to help knowledge management

Even though until now the focus has been AI and knowledge engineering methodologies, techniques, and technology as applied to knowledge management, the people and cultural aspects of knowledge management take precedence over the technology. A mantra within the knowledge management field is that 80% of knowledge management is people and cultural considerations while only 20% is technology. There are a number of lessons learned from developing and implementing AI solutions from an organizational perspective that can be applied to the KM field. This chapter examines some of these lessons.

Why should I give up my competitive edge?

In the construction of expert systems, some experts did not want to have their knowledge encoded in an expert system for others to use and share. They felt that their knowledge was their competitive edge, and they were not willing to give it up. Also, some people felt that if they used someone else's knowledge, then they could not put their own thumbprint on the knowledge acquired. To remove these barriers and concerns, many experts were willing to give up their knowledge for altruistic reasons as well as to allow them to free up their time to pursue other areas of interest. Many experts liked the recognition and the idea of having their knowledge preserved for eternity.

In many ways, knowledge management poses similar concerns as knowledge sharing vs. knowledge hoarding. People may not want to share their knowledge for competitive and self-preservation reasons. Incentives, such as being evaluated annually on knowledge sharing proficiency, are often provided to encourage knowledge sharing. Whether dealing with an AI or KM project, some motivation-and-reward system needs to be established to encourage the capture and sharing of expertise. Some tips for promoting knowledge sharing are (Stevens, 2000):

- Institute rewards and recognition programs to demonstrate knowledge sharing behaviors advocated by the company.
- Measure the success of knowledge sharing programs and use the metrics to gain support in the organization.
- Evaluate knowledge sharing behaviors as part of employee appraisals.
- Develop internal knowledge sharing leaders, whether a team or individuals.
- Form new teams or communities around subjects in which employees are interested.
- Involve current employees in the hiring process to maintain a team that works well together.

If you build it, they may not come

The information systems field has taught us that even if you build it, they may not come! Simply put, developing an AI or KM system and making it available to everyone does not necessarily mean that people will actually use it. This was apparent in the case of a successful software company whose chairman was a champion of knowledge management. The chairman was expecting new technology to produce a collaborative, sharing culture. It turned out that the company's greatest need was not new technology (which was strongly advocated by the company) but a culture modification program to prepare for a KM initiative (Barth, 2000). Having innovative technology solutions may be a piece of the puzzle, but implementation and user training plans will help ensure that people will respond to your efforts. The "throw it over the wall" approach, where users have to catch what is built for them, is problematic for successful use of the AI or KM system.

Resistance to change — make it part of the daily regimen

A number of consulting firms have created a knowledge portal to serve as a gateway to their daily computer activities as well as to their knowledge management system. For example, when a consultant comes into the office in the morning, he/she first enters the knowledge portal to check out any significant activities in his/her industry area, peruse various information and knowledge sources relevant to his/her area of focus, access email from within the portal, locate experts using the expertise locator function (as part of the KM system) via the portal, and share lessons learned from his/her respective community of practice as part of the KM system. The key point here is to institutionalize the KM system so that it becomes part of the daily/routine activities of employees. Roberts-Witt (2000) warns those involved in developing knowledge portals and KM systems not to underestimate resistance. She feels that users should not only know about the portal/KM system but also have incentives to use it. If your portal is connecting customers or suppliers, incentives and training are necessary to get those parties used to dealing with your company in this new way (Roberts-Witt, 2000).

When establishing online communities of practice, care and nurture need to be properly conducted and managed. Eisenhart (2000) offers some tips:

- Make sure the community has a compelling reason to exist, such as solving a problem common to all participants.
- Put company resources into communities aligned with company needs, such as choosing appropriate expansion sites or developing an improved product.
- Make it easy for members to participate.
- Use familiar technology. An expensive knowledge retrieval system is worthless if everyone is too busy to learn how to use it.
- Suit the technology to the task. Know when it is more effective to make a phone call than to hold a meeting.
- Make it easy for people to enter and leave the community as their responsibilities and skills are needed.
- Check with participants between formal gatherings and presentations to ensure that they are on track and have the resources they need.
- Help participants understand what tasks they are responsible for.
- Keep facilitators aware of each community's efforts as they relate to the larger entity and to other communities within the company.
- Create liaisons between the constituencies.
- Make it easy for future participants to find not just stored data, but any person who is likely to have a deep, dynamic understanding of the problem.

Make sure that the KM goals are aligned with corporate strategy

Many AI projects fail because the problems they solved do not contribute significantly to the mission of the organization. In a similar fashion, KM strategy and related KM initiatives must be integrated within the strategic vision of the organization and be closely aligned with corporate goals. If misalignment occurs, then the chances of success for the KM project are greatly threatened. At the U.S. Naval Sea Systems (NAVSEA) Command, one of the new seven strategic goals is knowledge management. In developing a KM strategy and program plan for the Naval Surface Warfare Center–Carderock Division (NSWCCD) (a subsidiary of NAVSEA), the KM strategy was developed to be in concert with NAVSEA's enterprise-wide KM strategy (although, in reality, the NSWCCD helped to establish the corporate-wide KM strategy for NAVSEA). Part of the knowledge management plan developed for NSWCCD is described below.

Knowledge management program plan

To help develop the knowledge management program plan for NSWCCD, a multicriteria decision-making methodology was used via Equity Software, a decision analysis tool used numerous times throughout the Division. Dennis Clark, Glenn Dura, and Jay Liebowitz developed the model as well as the relative weighting of the model.

The main components of the knowledge management plan included an expertise map, communities of practice with lessons learned capability, an expert and knowledge retention component, and an integrator of codified knowledge element. The expertise mapping linked knowledge areas, as developed in the knowledge taxonomy, with subject matter experts. The communities of practice, groups of individuals with shared trusts, beliefs, and values with a specific technical focus, included lessons learned/best practices component. The expert and knowledge retention element included the ability to capture, share, and retrieve key thought-making/decision-making processes relating to specific well-defined domains of knowledge. The integrator of codified knowledge relates to the integration of explicit knowledge (e.g., technical reports, documents, manuals, drawings) stored in various knowledge management-related databases in the division into one key database.

For each of these capabilities, different options were developed and examined. Specifically, for the expertise mapping, the options were: none, Anteon's Einstein, Advanced Skills Program (ASP), ASP plus our knowledge area taxonomy being developed in Code 011, Tacit Knowledge Mail Plus System, and Dataware Technologies' Knowledge Management Suite. For the communities of practice, the options were: none, Exchange/Outlook 2000 with folders, LiveLink (a NAVSEA-mandated document management tool) with threaded discussions, CommuniSpace, and Notes/Domino/Raven. For knowledge retention, the alternatives examined were: none, transcripts with hyperlinks, Sandia Labs' Video System (knowledge preservation project), Virage/CNN audio streaming, and a Web-based expert system. For the integrator of codified knowledge, the alternatives were: none, LiveLink, TIC (Technical Information Center) database enhancement, database for the technical stewardship program, and Dataware Technologies' Knowledge Management Server.

Each of these alternatives was weighted according to costs and benefits criteria. Cost criteria included: total costs derived from two-year estimates, cultural impacts associated with a change of "doing business" (e.g., more knowledge sharing vs. knowledge hoarding), and time to implement (amount of time to develop and implement the knowledge management option). Benefit criteria included: productivity associated with improved effectiveness and efficiency in meeting the needs of the "fleet of tomorrow"; innovation associated with the generation of new products or services and learning; attraction/retention referring to people retention, increased morale, sense of belonging, and knowledge retention; and return on vision as related to value-added benefits in meeting the strategic goals and vision of the organization to become the U.S. Maritime Technology Center.

After running the model and examining the portfolio of options along the frontier, the recommended proposed package was:

Expertise Map: Anteon's Einstein
Communities of Practice: Exchange/Outlook 2000 with folders
Knowledge Retention: Virage/CNN audio streaming
Integration of Codified Knowledge: TIC database enhancement (assumes the use of LiveLink)

Two other options deserve mentioning from our examination of the various portfolio mixes looking at benefits/costs. Using CommuniSpace as an alternative to Outlook 2000 should be considered as a mechanism for building communities of practice with a lessons learned (DISTILL function) component. Additionally, if Code 70 is greatly interested in preserving the thought-making processes involved with test and research and design methodologies, a Web-based expert system approach (using Exsys Web Runtime) could be explored.

Operationalizing the knowledge management plan

A knowledge portal should be created on NSWCCD's intranet/Web site to be the entry point to the various knowledge management capabilities, including the expertise locator, communities of practice with lessons learned, expert and knowledge retention, and (eventually) technical products/stewardship integrated database with links to KM-related databases such as Patents, Athena, etc. A knowledge management office should also be created with full-time staff to direct, oversee, help develop, and market knowledge management initiatives throughout this division and the Navy, as well as to educate division personnel about knowledge management.

In the first year of the knowledge management effort, it is recommended that a series of six-month pilots be established with a formal one-month measurement/evaluation phase. Since Code 70 is very interested in moving forward with knowledge management, a pilot should be set up with Code 70 to use Anteon's Einstein coupled with Outlook 2000 for the approximately 500 people in Code 70. The focus will be on providing a knowledge-sharing environment through an expertise mapping function, sharing information and knowledge, and building/nurturing a community of practice with a lessons learned component. It may also make sense to run a pilot of the CommuniSpace software across codes in an area such as advanced concepts to get a comparison of the usefulness and desirability of the CommuniSpace approach vs. the Einstein-Outlook 2000 approach. Separately, Code 70 may want to explore an expert/knowledge retention pilot project using the Virage/CNN audio streaming approach or a Web-based expert system approach for capturing the knowledge and experience of a key expert in a critical knowledge area. This could also be a six-month pilot with a one-month formal measurement/evaluation stage to follow. Simultaneously, the Technical Information Center should consider integrating their database into LiveLink for capturing, storing, and retrieving explicit codified knowledge such as technical reports and documents.

In the second year of the knowledge management initiatives, the plan calls for extending the online communities and collaboration effort to five more communities of practice throughout the division. We would expand the lessons learned repository to all directorates and continue to develop expert and knowledge retention systems in ten other areas.

People and cultural considerations will typically "make or break" a solution for solving the business problem. The applied AI community has learned these lessons over many years, as have personnel in the information systems field. Knowledge management is no different in this regard. Careful planning for building a supportive culture for knowledge sharing is critical for the success of a KM program. If the people, organizational, and cultural dimensions are fully considered up front, the KM strategy is likely to succeed.

References

Barth, S. (2000), "KM Horror Stories," *Knowledge Management Magazine*, October.
Eisenhart, M. (2000), "Around the Virtual Water Cooler," *Knowledge Management Magazine*, October.
Roberts-Witt, S. (2000), "Portal Pitfalls," *Knowledge Management Magazine*, October.
Stevens, L. (2000), "Incentives for Sharing," *Knowledge Management Magazine*, October.

chapter nine

Implementing knowledge management strategies

There will be a variety of knowledge management (KM) developments in the coming years that artificial intelligence (AI) technology will greatly impact. Wiig (1999) points out a few of these areas:

- There will be major knowledge transformation functions and repositories, such as capture and codification functions, and computer-based knowledge functions, such as training and educational programs, expert networks, and knowledge-based systems, and the different knowledge application or value-realization functions where work is performed or knowledge assets are sold, leased, or licensed.
- KM will be supported by many AI developments including intelligent agents, natural language understanding and processing, and knowledge representations and ontologies that will continue to develop and, by providing greater capabilities, will be relied on to organize knowledge and to facilitate knowledge application to important situations.
- Intelligent agents will not only acquire desired and relevant information and knowledge, but will also reason with it relative to the situation at hand (see Appendix D).
- An enterprise should experience faster organizational and personal learning through more effective discovery of knowledge via knowledge discovery and other systematic methods.
- There should be less loss of knowledge through attrition or personnel reassignments achieved by effective capture of routine and operational knowledge from departing personnel.
- Intelligent agents deployed internally and externally will offload data detective work required to locate and evaluate information required in many knowledge worker situations ranging from plant operators to ad hoc strategic task forces.
- Emerging sources of prepackaged knowledge will be sold (for example, LearnerFirst, Inc. is already doing this).

- Electronic advisory or consulting services are already emerging whereby knowledge-based systems can be bought in areas ranging from tax advice for individuals to water treatment for thermal power plants.

A tool to evaluate knowledge management strategies

Expert Choice 2000, by Expert Choice Inc. (www.expertchoice.com), is a decision support system that automates a process known as the analytic hierarchy process (AHP), developed by Thomas Saaty. It has been used extensively throughout the world for many years and could be helpful in evaluating which knowledge management strategy an organization should pursue. This tool and associated methodology was used by the author to develop a comparative analysis to determine which part of a knowledge management strategy should be pursued first. Expert Choice, like many decision support systems, is capable of supporting decisions with multiple criteria and alternatives. It allows the decision maker to incorporate his/her subjective as well as objective factors in the decision-making process. It does not have an explanation facility that is comparable to expert systems. However, sensitivity analysis can be performed using Expert Choice to perform numerical "what if" scenarios. Expert Choice is intended to be used in problems involving complex decisions — problems that entail multiple alternatives and several, often competing, criteria. Expert Choice represents a significant contribution to the decision-making process, as it is able to quantify subjective judgments in complex decision-making environments.

Expert Choice enables decision makers to visually structure a multi-faceted problem in the form of a hierarchy. At the top level (level 0), the goal is defined, such as to determine the most appropriate knowledge management strategy. At the next level(s), the criteria (or objectives) used in determining the knowledge management strategy are listed by the user. Then, the lowest level of the hierarchy lists the possible alternatives — in this case, the various knowledge management strategies under consideration. The criteria are:

Cost — the total cost of the knowledge management alternative

Culture — the cultural impact of the organization in terms of acceptance of the knowledge management approach

Time to implement — the amount of time needed to implement the KM approach

Productivity — the estimated worker productivity enhancements due to the KM approach

Knowledge retention — the ability to retain critical knowledge at high risk of being lost

People retention — the likelihood of retaining persons due to the KM approach

Return on vision — the return on the strategic vision due to the KM approach

After constructing this hierarchy, the evaluation process begins in which Expert Choice will first ask the user questions in order to assign priorities (i.e., weights) to the criteria. Expert Choice allows the user to provide his/her judgments in a verbal mode through pairwise comparisons so that no numerical guesses are required (it also allows the user to answer in numerical and graphical modes). Again, the user is responsible for determining the criteria and alternatives for his/her decision. In this analysis, the author acted as the user in supplying the judgments based on his experience. According to the user's verbal judgments, Expert Choice will calculate the relative importance through a scale ranging from 1 to 9, where:

1 is equal importance.
3 is moderate importance of one over another.
5 is strong importance.
7 is very strong importance.
9 is extreme importance.
2, 4, 6, and 8 are intermediate values between the two adjacent
    judgments.

This procedure is followed to obtain relative priorities of the criteria in which eigenvalues are calculated based on pairwise comparisons of one criterion versus another (see Figures 9.1–9.3). The criteria were ranked as follows after performing the pairwise comparisons:

Cost:	.126
Culture:	.087
Time:	.042
Productivity:	.115
Knowledge retention:	.206
People retention:	.236
Return on vision	.189

Figure 9.1 Weighted criteria after pairwise comparisons using Expert Choice.

Figure 9.2   Ranking of alternatives using Expert Choice.

Upon obtaining relative weights of the criteria, the alternatives are then weighted with respect to each criterion. The alternatives are:

- Expertise locator — an online yellow pages of experience mapping experts to knowledge areas
- Communities of practice/lessons learned repository — using group-ware to build and nurture online communities with shared interests and developing an associated lessons learned respository
- Web-based expert and knowledge retention system — a Web-based system that captures the decision-making processes of selected key experts

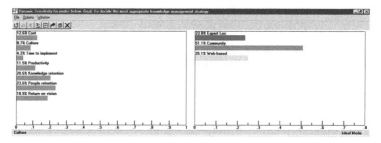

Figure 9.3   Another view after synthesis step.

After all the pairwise comparisons have been entered, Expert Choice performs a synthesis of adding the global priorities (global priorities indicate the contribution to the overall goal) at each level of the tree hierarchy. The end result of this synthesis is a ranking of the alternatives. An inconsistency index is calculated after each set of pairwise comparisons to show to what extent the user's judgments are consistent. An overall inconsistency index is calculated at the end of the synthesis as well. This measure is zero when all judgments are perfectly consistent with one another, and becomes larger when the inconsistency is greater. Saaty suggests that inconsistency is tolerable if it is 0.10 or less.

According to the synthesis step, the alternatives were ranked as follows:

1. Communities of practice/lessons learned repository   (.511)
2. Web-based expert and knowledge retention systems   (.251)
3. Expertise locator   (.238)

This suggests that the communities of practice/lessons learned repository would be the best approach to knowledge management for this particular example. The overall inconsistency index is .10, which is considered to be in the tolerable range.

Lessons learned from implementing expert and AI systems

A critical success factor for expert systems implementation and institutionalization is to have a champion in senior management who serves as the strong advocate for the system and change process. This is also true for knowledge management efforts, as senior leadership must actively support and use the knowledge management systems to serve as role models and for the knowledge management system to succeed. One chief executive officer of a leading organization would send e-mails to employees on a weekly basis if it appeared that they were not using the organization's knowledge management system (KMS). Additionally, if top management is using the KMS on a very frequent or even daily basis, then the employees throughout the company will feel that it must be important, so they will log on as well. Again, incentives such as frequent flier mileage for the first 500 users of the KMS and other creative enticements have been used by organizations to encourage initial use of the KMS.

The knowledge management strategy and resulting systems must be in alignment with the strategic vision of the organization and must be a well-designed program plan. Earlier chapters gave examples of knowledge management strategies and program plans. Similarly, expert systems technology has taught us these lessons. If the expert system does not contribute significantly to the "bottom line" or to the business strategy of the organization, then there is a lower likelihood of success for expert system usage and acceptance.

A last key lesson learned from expert system implementation is that most of these expert systems still support the decision maker instead of replacing him/her. The user still makes the final decision, and expert and other intelligent systems serve to help the user provide a structured framework for decision making. Similarly, the knowledge management system will not replace individuals — the KMS is there merely to facilitate interaction, collaboration, creation, sharing, and dissemination of knowledge.

References

Lamont, J. (2000), "Expert Systems and KM are a Natural Team," *KMWorld Magazine*, October.

Wiig, K.M. (1999), "What Future Knowledge Management Users May Expect," *Journal of Knowledge Management*, Vol. 3, No. 2.

chapter ten

Expert systems and AI: integral parts of knowledge management

Knowledge management generally consists of four functions: securing, creating, retrieving/combining, and distributing knowledge. Much of knowledge management is not new; its roots can be found in the expert systems and artificial intelligence fields. For example, the knowledge acquisition phase of expert systems can be applied to the capturing and securing of knowledge. Developing knowledge repositories for knowledge management activities can be easily traced to knowledge representation and knowledge encoding methodologies and techniques in the expert systems field. The indexing of knowledge can be traced to case retrieval, similarity, and adaptation methods applied in the case-based reasoning area of the expert systems field. Thus, many of the underpinnings of knowledge management are derived from earlier work in the expert systems and artificial intelligence field.

Even more important than acknowledging the roots of knowledge management in the expert systems field is the realization and understanding that expert systems should remain an integral part of a knowledge management system. Capturing expertise and putting it on-line in terms of on-line pools of expertise or Web-based interactive knowledge centers are actions critical to the potential success of knowledge management. For example, David Vandagriff (1997), Director of Technology Alliances for Lexis/Nexis, says, "We see it [Lexis/Nexis Exchange] becoming an on-line legal community that will include expert systems with Web-based engines using artificial intelligence. For instance, there will be a federal court venue expert. After a user responds electronically to a few questions, an answer will be given concerning proper venue." The U.S. Department of Labor (www.dol.gov/elaws) has already developed Web-based expert systems (e.g., determining veterans' benefits) as part of its knowledge compliance/knowledge management systems. Other companies and organizations are following suit to allow expert systems to play an important role in their knowledge management systems, but many others are still lagging behind.

Knowledge management involves understanding how the enterprise works. In an early article, William Stapko (1990) recognized that expert systems and other AI tools could greatly impact knowledge management activities. He stated, "Managing knowledge is a high-level corporate concern. Management wants to know how to run and manage a business using rules and guidelines to reference everything from marketing to manufacturing. Expert systems provide the ability to insulate the business knowledge from the technical knowledge."

The importance of expert systems in knowledge management

Expert systems and other artificial intelligence technologies have been maturing over the years. According to Hedberg (1995), AI may be hiding in many obscure places, but it is alive and kicking. Mrs. Fields Cookies, the Disney Store, the IRS, Microsoft, the White House, Xerox, Compaq, and many other organizations have used expert systems to assist them in their activities.

So the question remains, why don't more knowledge management officers recognize the importance and need for expert systems within their knowledge management structure? One director of knowledge management at a well-known organization in Washington, D.C. failed to see the significance of using expert systems and their underlying methodologies for his company-wide knowledge management effort. He felt that expert systems did not work in his organization when introduced years previously and his company preferred not to use expert systems.

I explained to this director that he is missing the boat and that expert systems technology (when applied to the appropriate problems and when expectations are not oversold) has matured to where it is a critical technology and business solution for many organizations; however, he did not seem convinced. That director and many other chief knowledge officers, or those with similar titles, may be underestimating the worth of expert systems usage within an enterprise knowledge management structure, especially when it comes to capturing the business rules of the organization.

Dertouzos (1997) mentions that another probable organizational development is the evolution of expert centers staffed by groups of related experts capable of high-quality, high-speed work at very competitive prices. Expert systems have a great role to play here. In fact, using expert systems as these expert centers for on-line expertise and help should be incorporated into knowledge management organizational systems, and will be, if knowledge managers realize the true potential of expert systems.

Liebowitz and Beckman (1998) describe an eight-step process for knowledge management. The stages are:

1. Identify — Determine core competencies, sourcing strategy, and knowledge domains.
2. Capture — Formalize existing knowledge.
3. Select — Assess knowledge relevance, value, and accuracy; resolve conflicting knowledge.
4. Store — Represent corporate memory in the knowledge repository with various knowledge schemata.
5. Share — Distribute knowledge automatically to users based on interest and work; collaborate on knowledge work through virtual teams.
6. Apply — Retrieve and use knowledge in making decisions, solving problems, automating or supporting work, job aids, and training.
7. Create — Discover new knowledge through research, experimentation, and creative thinking.
8. Sell — Develop and market new knowledge-based products and services.

Within this framework, expert systems could be used in the store, apply, and sell stages. According to Tom Beckman of the Internal Revenue Service, the field of AI is instrumental in many of these innovations. Value-added business comes from identifying and applying expert systems in situations where expertise and knowledge are required to solve problems. Knowledge engineers elicit expertise from domain experts and organize and structure it in ways that can be stored and applied in active forms to structure, guide, perform, and manage tasks, solve problems, and make decisions. The AI disciplines, and especially expert systems, can support the knowledge management process.

Expert systems can also be used as the integrative element linking various knowledge sources. They can serve as the integrative mechanism for solving interdisciplinary problems. Expert and knowledge-based systems provide the framework for handling the exchange and integration of knowledge from various sources. They allow knowledge bases to be created for ultimate sharing and analysis. They are an ideal technology for capturing, preserving, and documenting knowledge, especially in today's environment where organizations are reengineering, downsizing, and losing senior managers due to early retirement packages. Expert systems can be very useful for building the institutional memory of the organization before this intellectual capital is lost.

According to O'Leary (1997), shared knowledge is at the core of organizational or group memory and is essential to the preservation of expertise or process knowledge. DARPA's (Defense Advanced Research Projects Agency) intelligent information services project has moved to support virtual groups with a number of emerging technologies including: institutional memory tools that help organizations capture expertise, such as process knowledge and access to expert consultants; tools to support multiuser/multiauthor hypermedia Web development so groups can build their own Web sites; and self-organizing knowledge repositories that adapt to community needs with use. As an offshoot of expert systems, knowledge-sharing agents are starting to emerge, which could facilitate the knowledge management process.

Knowledge-based and expert systems have been noted for having the potential to play a major role in the knowledge management era. Karl Wiig of the Knowledge Research Institute in Arlington, Texas points out that historically the major impact of knowledge-based system applications in support of knowledge management has been to deliver knowledge to the point-of-action — where the most accurate information on the situation normally is present, analysis is performed, decisions are made, and the opportunity to serve the business in a timely manner is best (Wiig, 1997). However, at the present time, an increasing number of knowledge-based system applications can take on other roles such as building and organizing knowledge and supporting education, as well as many other purposes.

Holsapple and Joshi (1997) point out that expert systems can play a role in the framework of a knowledge resources component, a knowledge management activities component, and a knowledge management influences component. Certainly, expert systems could aid in the knowledge management activities component by representing and processing knowledge.

So what is the message here? The key message is that expert and knowledge-based systems and knowledge engineering should be recognized by knowledge managers as playing a fundamental role in the development of the organization's knowledge management system. Knowledge management is not new. Knowledge managers and knowledge analysts must lift up the outer lining of the knowledge management coat and examine what is underneath. They will quickly find that many of the methodologies, techniques, concepts, and tools from the expert systems and AI field can be appropriately applied to knowledge management. Knowledge management is the currently fashionable term for repackaging many of the ideas which developed from the AI/information technology, organizational behavior, and human resource management disciplines. As knowledge management philosophy expounds, we should learn from the past and these other disciplines so that we do not reinvent the wheel. One way of doing this is to apply expert systems/AI technology to capture these lessons and further apply this knowledge in a proactive manner. Use the power of expert systems technology in the knowledge management field, and there will be a greater likelihood for success of the knowledge management era (Liebowitz and Wilcox, 1997).

Closing thoughts for applying expert systems technology to knowledge management

Over the years, expert systems have contained procedural knowledge, declarative knowledge, episodic knowledge, and/or metaknowledge. These how to, what is, case-based, and knowledge-about-knowledge types of information can form the underpinnings for the knowledge repositories and corporate memories being built in knowledge management systems. Knowledge elicitation techniques borrowed from the knowledge acquisition community, as part of the expert systems field, can be applied to the securing and capturing phase of knowledge management. Knowledge sharing interchange formats, such as KIF, FIPA, and KQML, have already been developed as an outgrowth of the expert systems field to allow the sharing of knowledge to take place. Expert and knowledge-based systems can facilitate the collection of knowledge, and, when coupled with the connectivity of knowledge, the knowledge management system becomes a viable entity. Let us not reinvent the wheel — let us utilize the techniques that have worked so well for the expert systems and AI fields over the last two decades and apply the lessons learned in the expert systems and AI fields to the emerging field of knowledge management.

References

Dertouzos, M. (1997), *What Will Be: How the New World of Information Will Change Our Lives*, McGraw-Hill, New York.

Hedberg, S. (1995), "Where's AI Hiding?," *AI Expert*, CA, April.

Holsapple, C. and K. Joshi (1997), "Knowledge Management: A Three-Fold Framework," Kentucky Initiative for Knowledge Management, Paper 104.

Liebowitz, J. and T. Beckman (1998), *Knowledge Organizations: What Every Manager Should Know*, CRC Press, Boca Raton, FL.

Liebowitz, J. and L. Wilcox, Eds. (1997), *Knowledge Management and Its Integrative Elements*, CRC Press, Boca Raton, FL.

O'Leary, D. (1997), "The Internet, Intranets, and the AI Renaissance," *IEEE Computer*, January.

Stapko, W. (1990), "Knowledge Management: A Fit with Expert Tools," *Software Management*, November.

Vandagriff, D. (1997), *ABA Journal*, November.

Wiig, K. (1997), *Expert Systems with Applications International Journal*, Vol. 13, No. 1.

appendix A

A knowledge management strategy for the U.S. Federal Communications Commission*

Jay Liebowitz and Kent R. Nilsson

Summary

One of the major challenges facing the FCC concerns how best to capture, share, and leverage knowledge internally (and, in some respects, externally). Through early retirements, attrition, and an aging FCC workforce, the FCC is losing its key intellectual capital. According to MicroStrategy, Inc., 51 percent of all federal government employees are eligible to retire in 2001 (about 11% at the FCC, keeping in mind that many people left the FCC before it moved to the Portals). In order to combat this "brain drain," the FCC needs to develop a knowledge management strategy to capture, share, and preserve knowledge and integrate this strategy into their new strategic plan.

Knowledge management (KM) is the process of capturing institutional knowledge in ways that facilitate immediate and broader institutional use and ensure that information will be available in the future. The knowledge management services industry in the United States is currently a $1.3 billion business. According to Dr. Shereen Remez, the chief knowledge officer at GSA., about 80% of the Fortune 500 companies have knowledge management teams. Many federal government agencies and departments (e.g., NSA, NASA, SSA, HCFA, and DOL) have recognized the need for formal knowledge management in their organizations. GSA hired Dr. Remez, the first ever

* The authors gratefully appreciate the help of Mary Beth Richards (Deputy Managing Director, FCC), Dr. Stagg Newman (former FCC Chief Technologist), Matt Imhoff, the reviewers, and study participants.

chief knowledge officer in the U.S. government, to spearhead GSA's and the U.S. government's efforts in knowledge management. She is chairing a new subcommittee on knowledge management through the federal CIO (chief information officer) council.

If the FCC does not develop an effective knowledge management strategy and associated knowledge management systems, they will be in an increasingly precarious position. The Office of Engineering and Technology, with the help of Dr. Stagg Newman, Dr. Kent Nilsson, and Dr. Jay Liebowitz (the IEEE–USA FCC Executive Fellow), has been working with Mary Beth Richards, FCC deputy managing director, to develop a knowledge management strategy to be aligned with the FCC's new strategic plan. A knowledge management survey was drafted and pretested, and then distributed by e-mail to managers and staff in the recommended SUPMAN, OET, OPP, and ECONFCC listserves. The responses were compiled and analyzed, and follow-up discussions were conducted to probe deeper into key issues. The strategies and tools that are developed herein focus on creating more unified and integrated knowledge sharing networks; formalizing and systematizing knowledge capture, and building a supportive culture for knowledge sharing that includes strengthening incentives for knowledge reuse.

A quick overview of knowledge management

Many private sector organizations (e.g., advertising agencies, consulting firms, law firms, and investment banking firms) have long realized that their true strength (and value) lies in the brainpower or intellectual capital of their employees. This would appear to be particularly true for a government agency, such as the FCC, where no physical products are produced and where the FCC's contribution to society is almost wholly dependent upon the quality of the agency's thinking and the depth of knowledge that can be accessed in arriving at decisions. In order to remain effective and innovative in a world of accelerating change, organizations must learn to leverage their knowledge internally and externally. Web-based and intranet technologies technically enhance connectivity and help facilitate the sharing of knowledge. They also facilitate building bridges among isolated islands of knowledge.

Knowledge management involves more than effectively deploying technical delivery systems (i.e., computers, databases, and communications networks) for processing, storing, and sharing knowledge. It also involves developing a supportive culture with appropriate incentives. Building a supportive culture is a key element in any knowledge management strategy. Some organizations provide incentives to promote this until it becomes the norm. Other organizations require their employees to contribute actively to, and utilize knowledge in, the organization's knowledge repositories as part of their annual job performance evaluations.

Many organizations have found that both codification and personalization approaches are needed for an effective knowledge management strategy.

Codification refers to formalizing tacit knowledge that is typically difficult to express or explain by developing processes or mechanisms that permit this knowledge to become explicit and then become documented in a knowledge repository. Personalization refers to the sharing of ideas and tacit knowledge in informal settings. Knowledge management efforts can be very successful, but those that have failed can attribute their failure to either: (1) lack of senior management commitment and involvement, or (2) a poorly designed, or ill-conceived, knowledge management plan. To safeguard against these possible problems, the FCC's knowledge management strategy should be reviewed and supported by senior management.

Methodology for developing an FCC-wide knowledge management

Pretesting the knowledge management survey

A knowledge management survey was developed by looking at knowledge management receptivity, strategy/approaches/processes, culture, technology resources, and applications within the FCC. This survey (the revised version is provided at the end of this appendix) was pretested with the Network Technology Division (NTD) within OET before it was distributed by the SUPMAN, OPP, OET, and ECONFCC e-mail groups throughout the FCC. Some general NTD findings based upon the piloted knowledge management survey were:

- People are recognized as being the most important sources of information.
- Half the respondents felt that the FCC's strategic goals include knowledge management explicitly, and half the respondents felt there was a KM initiative at the FCC.
- Those who said that there was no KM initiative felt the idea had never been considered or discussed.
- Those who said that there was a KM initiative felt it was less than one year old.
- The top advantages of a KM initiative were perceived as being: (1) standardization of existing knowledge in the form of procedures/protocols and (2) facilitation of the re-use and consolidation of knowledge about FCC operations.
- The main approaches used to improve knowledge assets and knowledge sharing are cross-functional teams, communities of practice, intranet, and documentation/newsletters.
- The main approach for improving creation and refinement of knowledge is "lessons learned analyses."
- The main approach for storing knowledge is via regulations with FCC orders, cover memoranda, internal discussion/white papers, etc.

- The key knowledge that may be lost at the FCC is knowledge of non-published considerations behind the opinions (i.e., history of policy/implementation reasons for decisions).
- Most of the respondents felt that the top area for targeting knowledge management activities is competition/policy (i.e., institutional knowledge of underlying reasons for specific decisions), spectrum policy/management and related engineering knowledge, technical knowledge (in particular, synthesis of underlying knowledge of interconnection practices, regulations, and policy), knowledge of industry practices, and broad synthesis of all provisions of the Communications Act (especially Title III Radio and hearing provisions).
- The aspects of the organizational culture that support effective KM are the desire to reach out and provide expertise to other organizations and improved information flows via Web site, e-mail, the intranet, databases, etc.
- The potential inhibitors to KM are time pressures, high turnover of personnel, and usual turf protection.
- There is typically little or no organizational buy-in (currently) about KM among the staff and management, although there is a well-accepted "knowledge management system" that captures the legal logic and different stakeholder positions in the public records of FCC proceedings.
- There are no formal training programs to support knowledge management; in some cases KM is supported by on-the-job training and mentoring programs.
- Typically, steps have not been taken to reward and motivate people to encourage a knowledge sharing environment.
- Most people regularly use or have access to the intranet and Internet, but typically do not have, or use, more advanced technologies such as knowledge-based systems, lessons learned databases, and decision support systems.

Suggestions for refining the knowledge management survey were made and then incorporated into the revised survey. This survey was then sent electronically to about 400 people in the SUPMAN, OET, OPP, and ECON-FCC e-mail groups, representing all levels of management and staff across the offices and bureaus of the FCC. Ninety-six surveys were completed, giving a response rate of about 25% (this met our expectation that roughly 100 surveys would be fully completed).

Knowledge management survey analysis

Various symptoms were identified by survey respondents that suggest a strong need for knowledge management. These include:

- Too busy putting out fires now to think ahead

- Frequent transition of the chairman, office, and bureau chiefs
- Too busy to chat informally with colleagues
- E-mail used extensively in lieu of face-to-face discussions
- Valuable expertise left the FCC due to better job offers and early retirements
- On-line FCC "yellow pages" needed to map expertise and skill areas to individuals and groups within the FCC
- The government environment is transient in many professional areas, suggesting the need to capture valuable expertise before people leave
- Low training and development budget, which is essential to maintaining and replenishing human capital
- Overlooking the importance and historical context of FCC policy knowledge vs. operational knowledge

All of these symptoms and problems suggest a significant need to better share, leverage, and manage knowledge within the FCC.

The top areas for targeting knowledge management activities in the FCC were identified by the respondents as:

Competition/policy	27%
Enforcement	22%
Consumer information	21%
Licensing	13%
Convergence issues	13%
International communications	3%
Outreach to local/state governments	1%

The survey responses also provided additional information, listed in the tables below, and answers to the following questions:

Who are the most important knowledge carriers at the FCC?

People	56%
Paper	22%
Magnetic Media	13%
Processes	6%
Products/Services	1%
Other	2%

Do strategic goals explicitly include KM?

Yes	7%
No	93%

Is there a KM initiative in the organization?

Yes	16%
No	84%

Has the organization taken steps to motivate employees for KM initiatives?

Yes	17%
No	83%

How long has the KM initiative been in existence?

Less than 1 year	40%
1 to 2 years	5%
3 to 4 years	33%
More than 4 years	22%

Ranking of Advantages of KM at the FCC
(on a scale of 1 to 6, 1 being the most important;
multiple answers allowed)

Advantages of KM	1	2	3	4	5	6
Re-use and consolidation of knowledge about operations	24%	19%	18%	10%	13%	3%
Standardization of existing knowledge	26%	14%	14%	13%	6%	12%
Combination of customer knowledge and internal know-how	9%	12%	22%	13%	14%	15%
Acquisition of new knowledge from external sources	21%	12%	18%	10%	7%	15%
Generation of new knowledge inside the organization	17%	27%	15%	13%	8%	7%
Transforming individual knowledge to collective knowledge	50%	13%	8%	8%	3%	6%

Approaches Used to Improve Knowledge Assets

Approach	Organization-wide	Organizational-unit specific	Methods (no indication)
Sharing and Combination of Knowledge			
External or internal benchmarking	8%	10%	6%
Communities of practice	16%	18%	21%
Cross-functional teams	21%	16%	18%
Intranets (including groupware)	21%	5%	18%
Training and education	25%	15%	26%
Documentation and newsletters	20%	13%	22%
Other	1%	2%	2%
Creation and Refinement of Knowledge			
Lessons learned analysis	9%	11%	20%
Research and development centers/labs	2%	8%	4%
Explicit learning strategy	6%	5%	6%
Other	0%	1%	1%
Storing of Knowledge			
Storage of customer/ stakeholder knowledge	8%	10%	11%
Best practice inventories	7%	9%	3%
Lessons learned inventories	8%	13%	8%
Manuals and handbooks	19%	19%	8%
Yellow pages of expertise/ knowledge	7%	10%	0%

Degree of Organizational Buy-In for Knowledge Management

Employee	None	A little	Some	A lot
Staff				
Professionals/knowledge workers	19%	29%	22%	13%
Operational and clerical	28%	29%	17%	5%
Management				
Senior management	17%	24%	25%	13%
Middle management	17%	23%	25%	15%
First line supervisory	16%	24%	26%	10%

Information Technology used as an enabler to:

Investigate, assess, and safeguard important knowledge	72%
Use best knowledge to do job well	60%
Learn and innovate to do job better	57%
Reengineer workplace and production system	33%
Better inform public and constituents	68%
Create new products and services	23%
Other	1%

Technologies Used Regularly

Technology	Percent using
Intranet Technologies	
E-mail	94%
Videoconference	23%
Yellow pages of expertise	16%
Discussion forums	34%
Shared documents/products	54%
Training and education	44%
Gathering and publication of lessons learned/best practices	26%
Internet Technologies	
Knowledge searching on the World Wide Web	83%
Knowledge exchange with customers	47%
Knowledge and Database Technologies	
Knowledge-based systems (expert systems)	17%
BP/LL databases (best practices/lessons learned)	9%
PSS/DSS (performance/decision support systems)	15%
Management information systems (transaction processing)	42%
Data Mining and Knowledge Discovery Techniques	
Extracting knowledge from process data to improve operations	23%
Simulation	12%

The main findings from these responses were:

- People were consistently identified as the most important knowledge carriers; this is a good foundation for knowledge management activities and beliefs.
- Knowledge management is not an explicit part of the FCC's current strategic goals.
- The top advantage of knowledge management is to transform individual knowledge into organizational knowledge.

- Training and education were cited as the main approaches for sharing knowledge, yet there has been a very limited budget for formal training and education — especially for the FCC's field staff.
- Lessons learned analyses was the top method for creating and refining knowledge, yet codifying them into an online repository has rarely been done at the FCC.
- Manuals and handbooks seem to be the top way of storing operational knowledge at the FCC.
- There is "a little to some" buy-in at both the staff and management levels for the basic tenets of knowledge management.
- The FCC, however, has not taken steps to motivate or reward employees for creating and contributing to a knowledge sharing culture.
- Information technology is primarily used at the FCC to investigate, assess, and safeguard important knowledge, and to better inform the public and the FCC's constituents.
- Technologies used most frequently at the FCC include e-mail, knowledge searching on the Web and specialized databases (such as West-Law), and management information systems. Very little use occurs with respect to decision support tools (e.g., computerized multi-criteria decision-making tools), advanced information technologies (e.g., knowledge-based systems that capture and emulate the behavior of human experts in well-defined tasks), and best practices/lessons learned databases.

Findings and recommendations

In order to meet the goals of the FCC's new strategic plan, several knowledge management goals should be integrated into the FCC's strategic vision. These include:

- Further increasing and facilitating employee access to the knowledge needed to do the jobs well
- Further improving the quality and comfort level (e.g., reliability and impartiality) of FCC decisions
- Capturing and storing, to the extent possible, employee knowledge that is critical to the FCC's operations
- Instilling a culture of knowledge sharing and reuse within the FCC to facilitate education and mentoring in the years ahead

To achieve these goals, the following objectives need to be addressed (Berens, 1999):

- Create a more unified or integrated knowledge network. Most of the knowledge sharing within the FCC occurs informally, either between individuals in their work units or teams or between peers across work units or groups. This informal knowledge sharing personalization approach should be encouraged and continued, but the tacit knowledge exchanged during these meetings needs to be made explicit and codified into a formal knowledge repository. If knowledge is not shared, the following could happen (Cortada and Woods, 1999): repeated mistakes, dependence on a few key individuals, duplicated work, lack of sharing of good ideas, and slow introduction of new products or market solutions. Presently, there is no easy way to search across existing knowledge repositories or directories. A global FCC knowledge management system should be created with a uniform technology infrastructure. This knowledge management system should include proven practices (best and worst practices), success stories, lessons learned, and rules of thumb (heuristics) gained from experiential learning. Additionally, a need exists to create an organizational yellow pages of expertise to map knowledge/skill areas to individuals throughout the FCC.

- Formalize and systematize knowledge capture. The majority of the FCC's intellectual capital currently resides in the minds of its employees. Because of the high turnover rate in the FCC employee base, there is a great need to capture the knowledge of the FCC experts before the experts retire. Widely accessible and thorough documentation of core FCC competencies also needs to be created and maintained in some repository. The oral history of the nonpublished logic behind the opinions (i.e., history of policies, political process, reasons for decisions) may need to be captured. Since much of the knowledge is case-based in nature, a knowledge management system applying case-based reasoning may be useful. Case-based reasoning deals with looking at previous similar cases or scenarios to help answer a new situation which may have similar features.
- Strengthen incentives to re-use knowledge. Encouraging staff to re-apply successful practices and re-use available knowledge could improve the FCC's responsiveness (via the case-based knowledge management system approach described above). Forming communities of practice or online communities may also be useful to the FCC to encourage the elicitation, sharing, and transfer of tacit knowledge in the organization. Some organizations such as the World Bank have knowledge fairs where they encourage people with similar interests in the organization to network, share experiences, tell stories, and build a stronger community of practice. The FCC's practice of brown bag lunches for economists, MBAs, and new lawyers is an excellent example of how to share knowledge and build communities of practice. Additionally, as part of the annual job performance review, individuals could be evaluated on how well they are contributing to knowledge repositories and how well they are applying, in a value-added way, the knowledge retrieved from these repositories.

Organizationally, the FCC should consider hiring a chief knowledge officer (CKO) and associated knowledge management team who would report directly to the chairman or managing director. The CKO should not be the CIO (chief information officer) because a different skill set is needed and a strictly technology-focused approach to knowledge management should be avoided. There should also be knowledge managers and knowledge specialists throughout the various departments at the FCC. The knowledge manager's role would be to gather, review, and oversee/maintain the knowledge assets for a given area. The knowledge specialist may be a subject matter expert who is detailed for a certain amount of time to gather, analyze, and synthesize a body of knowledge critical to the FCC.

Culturally, the FCC may want to add a knowledge performance factor into the annual job performance reports. As appropriate, employees should be held accountable for contributing to the FCC's knowledge stores as well as sharing, reusing, seeking, and applying the knowledge from the knowledge repositories. Some organizations, like Andersen Consulting, even have knowledge sharing proficiency levels, and to be promoted, employees must reach a certain level. For each substantial FCC project, part of the process should be to document/encode a set of lessons learned from the project's undertaking into the knowledge management system. This should be standard practice as part of the evaluation of all substantial projects.

For knowledge management to work and to build a supportive culture for knowledge sharing, people must be recognized and rewarded for their work. Recognition could include associating each person's name with his/her respective lessons learned in the knowledge repositories and then publishing on the intranet, on a monthly basis, the most frequently accessed or reused nuggets of knowledge with the authors' names attached. American Management Systems uses this approach in its Best Knews headlines, and Xerox has used name recognition in associating people's names with the tips/lessons learned that they provide (this has been more motivating than cash incentives). The FCC's bureau/office weekly report to the chairman, which is shared with all bureau/office chiefs and selected others, is a wonderful vehicle for encouraging a knowledge sharing environment (perhaps this report should to be distributed more widely). By also incorporating a knowledge contribution and usage factor into an individual's annual job performance review, employees will be rewarded monetarily for their efforts. Job rotation may also be considered in order for employees to better understand the various FCC functions and to further stimulate them.

Next steps towards making the FCC a knowledge organization

The following are recommendations for the next steps that could be accomplished in the near future (within one year):

- Select a chief knowledge officer for the FCC (reporting directly to the chairman or managing director) to spearhead knowledge management initiatives, and establish a knowledge management team across the various offices and bureaus (including HR) in the FCC as a KM steering committee (the CKO would chair this committee).

- Establish a knowledge transfer department overseen by the CKO would oversee to help develop the FCC-wide knowledge management system, provide educational forums on knowledge management throughout the FCC, and act as the analyzer and disseminator of lessons learned to appropriate individuals within the FCC who would could benefit from these lessons.
- Create an on-line organizational yellow pages of expertise to map knowledge/skill areas to individuals throughout the FCC.
- Bring back some recently retired key experts in part-time annuitant positions to capture their knowledge for codification into the knowledge management system.
- Develop a certification/training curriculum for key FCC processes and functions.
- Increase the training and development budget and develop a formal mentoring program.
- Develop a plan to hire new staff members on a continual basis (e.g., re-institute the engineers-in-training program, law clerk training program, COOP program). The FCC needs to have new staff in the pipeline so when someone retires there is staff available to become the new leaders of the future.
- Each bureau/office should, on a routine basis, present tutorials on hot topics in their respective areas for all interested FCC employees, and these tutorials should be made available throughout the FCC.
- Make sure that each bureau/office has in-depth manuals of certain processes available via the intranet and updated regularly (this need was greatly lessened due to streamlining initiatives).
- Use the team concept on many issues so that people meet and learn from each other.
- Include a knowledge performance factor as part of the annual job performance review report and provide a reward structure to motivate employees to share knowledge.
- Make standard the practice that lessons learned must be encoded as part of the on-line knowledge management system each year during project development and implementation, before final sign-off can occur.
- Start to develop, via the FCC intranet and knowledge management tools, an on-line knowledge management system which will include best/worst practices, lessons learned, yellow pages of expertise, appropriate documentation, cases, and on-line pools of mission-critical FCC knowledge and expertise that are captured and accessible in an interactive way. The top areas for targeting knowledge management activities in the FCC are competition/policy, enforcement, and consumer information; the oral history of the nonpublished logic behind the opinions (i.e., history of policies, political process, reasons for

decisions) may need to be captured (as well as lessons learned), keeping in mind that there may be some oral histories that should not be documented for legal and other reasons.

- Capture the following areas of knowledge relating to competition/policy in a knowledge management system: institutional knowledge as to why and how some things were decided, historical price trends tracking, spectrum policy and management and associated engineering knowledge, technical knowledge such as interconnect regulations and policy, industry practice knowledge, knowledge of past legal precedents and historical political policy decisions, and broad overview of all fundamental provisions of the Communications Act (especially Title III radio and hearing provisions).
- Offer a knowledge exchange in key FCC core competencies as an informal get-together to allow communities of practice to network and share their knowledge.
- Make sure that every employee has the necessary hardware/software/telecom facilities to access the knowledge management system.

References

Berens, M. (1999), Knowledge Management Plan, American Association for Retired Persons (AARP), Washington, D.C.
Cortada, J. and J. Woods, Ed. (1999), *The Knowledge Management Yearbook 1999-2000*, Butterworth-Heinemann, Cambridge, MA.

Bibliography

Davenport, T. and L. Prusak (1998), *Working Knowledge*, Harvard Business School Press, Cambridge, MA.
Liebowitz, J. (2000), *Building Organizational Intelligence: A Knowledge Management Primer*, CRC Press, Boca Raton, FL.
Liebowitz, J., Ed. (1999), *The Knowledge Management Handbook*, CRC Press, Boca Raton, FL.
Liebowitz, J. and T. Beckman (1998), *Knowledge Organizations: What Every Manager Should Know*, CRC Press, Boca Raton, FL.

Knowledge management survey

In the context of this survey, we consider "Knowledge Management" to include the strategies, tactics, and support mechanisms for the creation, identification, collection, and sharing of knowledge and practices and applying the best knowledge within the organization. The purpose of knowledge management is to improve your organization's effectiveness by leveraging the knowledge you have and need to use and to build, capture, and preserve the intellectual capital ("knowledge base") of the organization. Knowledge typically refers to *how* and *why* answers to questions, whereas information typically answers *who*, *what*, *when*, and *where* questions.

This survey is being conducted under the sponsorship of Mary Beth Richards, Deputy Managing Director, by Dr. Jay Liebowitz, the IEEE–USA Executive Fellow at the FCC and the Robert W. Deutsch Distinguished Professor of Information Systems at UMBC, in collaboration with Dr. Stagg Newman, Dr. Kent Nilsson, and colleagues in the OET area. The main goal is to try to develop a knowledge management strategy and program plan for the Commission to hopefully be integrated within the new strategic plan of the FCC.

This survey will have several parts, composed partly of questions from the APQC (American Productivity and Quality Center), CIBIT (Center for Business Information Technology), and other organizations. Kindly respond as honestly as possible and return the completed survey to Dr. Jay Liebowitz, OET, Room 7-C120 by December 10, 1999. Thank you in advance for your help. A white paper will be written based on the analysis and results of this survey.

Strategy, approaches, and process

Who or what are the most important knowledge carriers in your organization?
_____People
_____Paper
_____Magnetic Media
_____Processes
_____Products /Services
_____Other (please specify)_____

Do your organization's overall strategic goals explicitly include knowledge management?
_____Yes _____No

Do you have a knowledge management initiative in your organization?
_____Yes ___No

If "no," what are the reasons why a knowledge management initiative may not exist?
_____
_____
_____

If "yes," how long have you had a knowledge management initiative?
_____less than 1 year
_____1–2 years
_____2–4 years
_____4 years or more
_____other (please specify)

What do you see as the advantages of a knowledge management initiative to the FCC? Please rate the following "knowledge management objectives" in the context of your organizational strategy with 1 = highest importance and 6 = lowest importance.

_____ Facilitation of the "re-use" and consolidation of knowledge about operations
_____ Standardization of existing knowledge in the form of procedures/ protocols
_____ Combination of customer knowledge and internal know-how
_____ Acquisition of new knowledge from external sources
_____ Generation of new knowledge inside the organization
_____ Transforming individual (people's) knowledge into collective knowledge (i.e., transforming individual learning into organizational learning)
_____ Other (please specify)_____

What approaches do you use to improve your knowledge assets and operations?
a) Sharing and combination of knowledge
__external or internal benchmarking
__communities of practice (expert groups)
__cross-functional teams
__intranets (including groupware)
__training and education
__documentation and newsletters
__other (please specify)_____

b) Creation and refinement of knowledge
__lessons learned analysis
__research and development centers/labs
__explicit learning strategy
__other (please specify)_____

c) Storing of knowledge
__storage of customer/stakeholder knowledge
__best practice inventories
__lessons learned inventories
__manuals and handbooks
__yellow pages of expertise/knowledge
__other (please specify)_____

For a,b, and c above, please put a "+" after those approaches which are organization-wide and a "−" after those approaches which are organizational unit-specific.

Are there specific knowledge management-related examples, activities, practices, functions, or capabilities that you consider to represent your best knowledge management efforts?

_____
_____

What skill sets and expertise are needed for the knowledge management-related examples above?

_____
_____

Name one or two major functions in the FCC in which you feel that critical/key knowledge is being lost due to attrition, early retirements, lack of methods for capturing expertise, etc.:

_____
_____
_____

What would be the top two areas for targeting knowledge management activities:
_____Enforcement
_____Consumer information
_____Licensing
_____Competition/policy
_____International communications
_____Convergence issues
_____Broadband access to rural areas
_____Other (please specify)_____

Culture

Which aspects of your organizational culture seem to support effective knowledge management?

_____
_____
_____

Which aspects of your culture seem to be barriers to effective knowledge management?

_____
_____
_____

To what extent has there been organizational buy-in and acceptance about knowledge management at the following levels?

Staff:

Professionals/knowledge workers___None ___A little ___Some ___A lot ___Total

Operational and clerical workers ___None ___A little ___Some ___A lot ___Total

Management:

Senior management   ___None ___A little ___Some ___A lot ___Total
Middle management   ___None ___A little ___Some ___A lot ___Total
First line supervisory ___None ___A little ___Some ___A lot ___Total

Do you have specific training programs in place to support knowledge management?

_____

_____

_____

Has your organization taken steps to motivate and reward people and/or teams supportive of effective knowledge management?
___Yes ___No

Please list or describe briefly any incentives/reward systems that support knowledge management.

_____

_____

_____

What insights or lessons learned have you experienced thus far with respect to initiatives (such as Total Quality Management) that may affect the culture of the organization?

_____

_____

_____

Technology

Do you use information technology (IT) as an enabler to (check all that apply):

_____Investigate, assess, safeguard important knowledge

_____Use the best knowledge to do the job well

_____Learn and innovate to do the job better

_____Reengineer the workplace and the production system

_____Better inform the public and your constituents

_____Create new products and services

_____Other (please specify)_____

Which of the following technologies do you regularly use (or have at your disposal to use) to support knowledge management initiatives:

Intranet (internal Internet based upon World Wide Web browsers or group-ware):

_____E-mail

_____Videoconferencing

_____Yellow pages of expertise (knowledge maps — linking areas of expertise with experts in the FCC)

_____Discussion forums

_____Shared documents/products

_____Training and education

_____Gathering and publication of lessons learned/best practices

Internet Functions:

_____Knowledge searching on the World Wide Web

_____Knowledge exchange with customers

Knowledge and Databases:

_____Knowledge-based systems (computer programs that emulate the behavior of experts in well-defined tasks)

_____Best practice/lessons learned databases

_____Performance support systems/decision support systems

_____Management information systems (traditional transaction processing systems)

Data Mining and Knowledge Discovery Techniques:

_____Extracting knowledge from process data to improve operations

_____Simulation (use of computer modeling for what-if scenarios)

Which technologies do you plan to use to support knowledge management?

_____

_____

_____

Please provide any other comments that would help us in developing a knowledge management strategy for the Commission:

_____

_____

THANK YOU AGAIN FOR SPENDING YOUR VALUABLE TIME COMPLETING THIS SURVEY.

appendix B*

Partial knowledge audit for the U.S. Social Security Administration

Knowledge management receptivity

In order to check the receptivity for knowledge management within those experts involved in the Social Security Administration's (SSA) benefit rate increase (BRI) process, a knowledge management survey was distributed, compiled, and analyzed. The knowledge management receptivity component of this survey is a validated instrument developed by Professor Nick Bontis, director of the Intellectual Capital Research Centre at McMaster University in Canada. The survey instrument has twelve questions segmented into three general parts — a set of questions dealing with the impression and image of the organization's experts, another set dealing with familiarity of knowledge management concepts and terms, and the last set dealing with developing a knowledge management infrastructure within the SSA.

Expert and knowledge workers

The survey indicates that 87% of the respondents either agreed or strongly agrees that managers need to be aware of the importance of providing their expert employees with challenging work to retain knowledge in organizations. Eighty-seven percent also indicated that expert employees are the most valuable resource for organizations. Perhaps coincidentally, 87% also indicated that organizations need to have clear strategies for retaining their expert employees.

---

* Adapted from Buchwalter, Rubenstein-Montano, and Liebowitz, 2000.

To retain knowledge in organizations, managers need to be aware of the importance of providing their expert employees with challenging work.

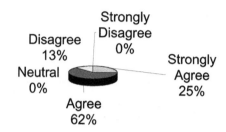

Expert employees are the most valuable resource for organizations.

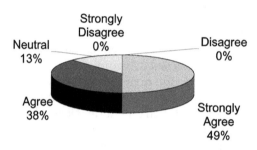

Organizations need to have clear strategies for retaining their expert employees.

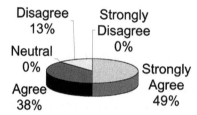

Knowledge management terms and concepts

Most people were unfamiliar with the term *knowledge work*. Fifty-eight percent were neutral, and 14% did not feel that higher level professional work would correctly be termed *knowledge work*. Seventy-four percent agreed or strongly agreed that knowledge workers are the primary contributors to success in organizations. Seventy-four percent responded that they were unfamiliar with "knowledge organizations" as a term and concept.

Higher level professional work would correctly be termed "knowledge work."

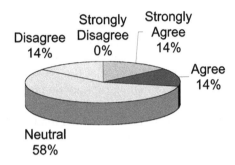

Knowledge workers are the primary contributors to success in organizations.

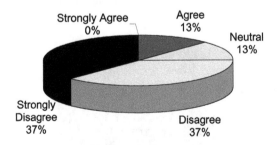

I am familiar with "knowledge organizations" as a term and concept.

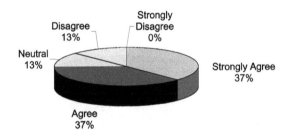

Knowledge management infrastructure

There seemed to be ambivalence (58%) and disagreement (28%) to the statement, "Our organization is ready to transform itself into a learning organization." This indicates that the SSA may have some distance to travel in order to transform itself into a learning organization and to convert individual learning into organizational learning. Seventy-five percent felt that a knowledge perspective is needed for today and in the future to be truly successful. In terms of designating a chief knowledge officer (CKO) at the SSA, 25% agreed that the position should be created, 62% were neutral, and 13% disagreed. It should be pointed out that GSA (Government Services Administration) recently hired a CKO for its organization. Additionally, an Arthur D. Little study, performed a few years previously, indicated that 41% of the Fortune 500 companies had a CKO or equivalent position in their organization. There was overwhelming agreement (87%) that the SSA should commit additional human/financial resources to managing knowledge. In terms of the SSA having the leadership capability to succeed in knowledge management, 57% agreed, 29% were neutral, and 14% disagreed. Finally, 50% felt it was important to measure intangibles in the organization, 25% were neutral, and 25% felt it was not necessary.

Our organization is ready to transform itself into a learning organization.

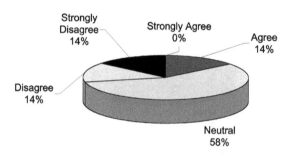

To be truly successful today and in the future, one needs to see the world from a knowledge perspective.

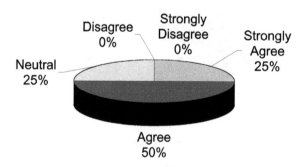

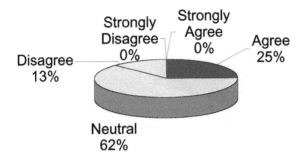

A distinct CKO position should be a part of this organization's structure

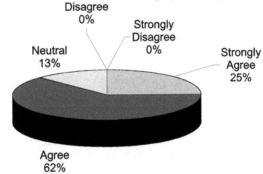

Our organization should commit additional human/financial resources to managing knowledge

Our organization has the leadership capability to succeed in knowledge management

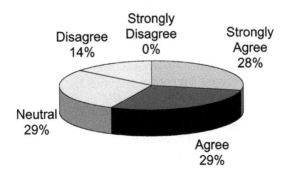

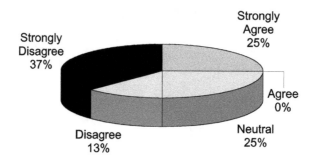

What is intangible in organizations is hardly worth measuring

- Strongly Agree 25%
- Agree 0%
- Neutral 25%
- Disagree 13%
- Strongly Disagree 37%

Knowledge management culture

Respondents indicated a lack of awareness and/or consensus regarding the cultural attributes of the SSA. The lack of awareness is illustrated by a large number of questions left blank or answered "don't know" This finding corresponds to the following questions:

Question	Number of blank or "don't know" responses
Which aspects of your organizational culture seem to support effective knowledge management?	7

To what extent has there been organizational buy-in and acceptance about knowledge management at the following levels?
    Management: senior management, middle management, supervisory management
    Staff: professionals/knowledge workers, operational and clerical workers
	4 or 5 for each (except 6 for operational and clerical)

Do you have specific training programs in place to support knowledge management?	1
Has your organization taken steps to motivate and reward people and/or teams supportive of effective knowledge management?	2
Please list or describe briefly any incentive/reward systems that support knowledge management.	8
What insights or lessons learned have you experienced thus far with respect to culture changes in your organization?[a]	7

[a] This may also have been due to the fact that respondents found the question confusing.

The lack of consensus is also supported by respondent answers. This conclusion is drawn primarily from responses to two of the questions: (1) "Which aspects of your organizational culture seem to support effective knowledge management?" and (2) "To what extent has there been organizational buy-in and acceptance about knowledge management at the following levels?"

For the first of those two questions, one respondent cited a variety of written documents such as manuals, functional requirements, and annotated system code. The same respondent also noted training programs. Another respondent answered along similar lines by mentioning formalized procedures. Then, a different respondent answered the same question by saying that there are no formal aspects of the SSA culture for supporting knowledge management because employees are too busy to concern themselves with mentoring and learning new things. Another point without consensus is the issue of resources. One respondent answered the first question by saying that there are additional team members "to insure the continuity of the BRI/PAC process." But the respondent who cited the lack of time for mentoring and learning also noted that there are no back-up analysts for some of the positions in the BRI/PAC area. Three respondents also cited insufficient resources as the key barrier to effective knowledge management in answering the question, "Which aspects of your culture seem to be barriers to effective knowledge management?"

Answers to the second question also indicated a lack of consensus. Answers to this question were as follows:

Level	Blank/don't know	None	A little	Some	A lot
Senior Management	4	3	2	—	2[a]
Middle Management	5	3	2	—	1
Supervisory Management	5	2	1	2	1
Professionals	4	1	3	2	1
Operational and clerical	6	—	3	1	1

[a] One "a lot" response was from a senior mgmt respondent who supports KM.

In addition to these two examples of lack of awareness and lack of consensus, constructive responses were received that correspond to the following questions: (1) "Which aspects of your culture seem to be barriers to effective knowledge management?" (2) "Please list or describe briefly any incentive/reward systems that support knowledge management," and (3) "What insights or lessons learned have you experienced thus far with respect to culture changes in your organization?"

As already mentioned, question 1 above cited resources as the primary barrier to knowledge management (3 responses). In addition, two respondents noted a lack of recognition of individuals as a barrier. The respondents expanded on this point to explain that individuals are not recognized for the knowledge they possess and team leaders are not assigned on merit and experience, but rather are rotated through different work areas. One respondent also mentioned division lines (i.e., hierarchy) in the SSA and upper management as barriers, and another noted that the SSA is driven by legislation and claimant/recipient needs.

Question 2 above is a follow-up to the question, "Has your organization taken steps to motivate and reward people and/or teams supportive of effective knowledge management?" To the first part of the question, seven respondents answered no and two answered yes. The yes responses then led to answers for our second question above regarding an awards program (one respondent even mentioned cash awards). However, this response is in direct conflict with another answer for the same question which stated that emphasis is on meeting deadlines regardless of how those deadlines are accomplished. Again, the individual, as well as the quality of his/her work, is unimportant as long as the job gets done.

Question 3 above confused several of the respondents. But for those who did provide an answer, the comments were fairly negative with responses such as, "Roll with the punches — don't let management fads get in the way of getting the job done"; "Individuals are not recognized for their knowledge. Management is much more concerned with promoting and awarding based on political concerns"; and "Most people resist changes." These answers indicate a general trend that the organization's culture is fairly negative and unfocused on individuals. Last, someone responded to this question by saying that documentation at the SSA is far from complete. Things such as coordination activities, validation techniques, validation tools (playback nuances) are not currently documented. The requirements area is, however, well documented.

Knowledge management technology

Virtually all respondents recognized the use of some form of computer technology to accomplish knowledge management tasks. Eighty percent of the respondents recognized that they use IT (information technology) as an enabler to their work. The following table lists the percentages of the respondents that use IT for these tasks.

Task	Percent using IT	Not understood
Investigate, assess, and safeguard important knowledge	30%	20%
Use the best knowledge to do the job well	50%	
Learn and innovate to do the job better	30%	
Reengineer the workplace and the production system	20%	
Better inform the public and your constituents	30%	
Create new products and services	40%	

There was broad recognition of intranet usage and applications among the respondents. Eighty percent were aware of a Webmaster at the SSA. The following table lists the percentage of respondents who were aware of various technologies that can be used to support knowledge management at the SSA.

Technologies	Percent using technologies
E-mail	100%
Videoconferencing	50%
Yellow pages of expertise (knowledge maps)	10%
Discussion forums	30%
Shared documents/products	70%
Training and education	90%
Gathering and publication of lessons learned/best practices	0%
Knowledge searching on World Wide Web	20%
Knowledge based systems	30%
Performance support systems/decision support systems	10%
Management information systems	80%
Extracting knowledge from process data to improve operations	20%

Although most people (80%) did not believe or know whether knowledge bases were accessible through the intranet, one person did list three applications — MI, shared documents, and training — that are available on the intranet. Again, most people (90%) did not believe or know whether any knowledge bases were going to be made accessible through the intranet. No one knew which technologies were going to be used for knowledge management.

Knowledge management outcomes

Most respondents were not measuring and tracking knowledge. One of the respondents did track knowledge and had the following comment: "I try to keep abreast of vulnerabilities resulting from potential resource losses. Also, we still push for backups, and at the end of projects I'm involved in doing process checks (lessons learned)."

Major benefits of KM	Percentage of responses
Increased innovation	20%
Practice and process improvement	60%
Increased customer satisfaction	30%
Enhanced employee capability and organizational learning	50%
Improved efficiencies in writing reports and responding to inquiries	10%
Lower learning gap	10%

A recurring comment from the respondents, found in about 30% of the responses, can be summarized in the following statement that was made by one of the respondents: "This is the first time I have ever heard the concept 'Knowledge Management.' I am not a member of management so I do not know if this concept is something that SSA is now working on." It was apparent from the responses that almost all of the respondents were not on the management level, which explains their lack of knowledge regarding future plans and other organization-wide initiatives.

References

Buchwalter, J., B. Rubenstein-Montano, and J. Liebowitz (2000), Benefit Rate Increase/Premium Amount Collectable Knowledge Management Study, University of Maryland–Baltimore County/Social Security Administration, White Paper, January.

appendix C

Modeling the intelligence analysis process for intelligent user agent development*

Joshua Phillips, Jay Liebowitz, and Kenneth Kisiel

The competitive intelligence model

The development of a comprehensive intelligence model which has been encoded into a tool called POINT (Problem Organization INtelligence Tool), developed by the authors in Visual Basic and MS-Access. The rest of this appendix details that model.

The next step was to further develop this agent by enabling it to monitor the user's actions and provide suggestions during the competitive intelligence process. In order to provide this level of intelligence, the knowledge model of a typical intelligence operation had to be developed.

Both interviews and a literature review were used in order to create the model. Typical knowledge elicitation and modeling methodologies were only partially applicable because the problem domain was so abstract. However, an attempt was made to follow the general form of the task analysis worksheet that is part of the CommonKADS methodology (Schreiber et al., 2000).

For each task that was identified within the larger analysis process, we explored the following aspects:

* We gratefully acknowledge the support of the Maryland Industrial Partnership System (MIPS) grant and WisdomBuilder, Inc. for funding this effort.
Adapted from the *Research and Practice in Human Resources Management Journal*, National University of Singapore, 2001. With permission.

- the goal of the task
- the inputs and outputs
- the structures that we manipulated by the task
- pre-conditions and post-conditions
- the agents involved
- factors for judging successful completion of the task

The literature review was a more practical method than interviewing or eliciting knowledge about such an abstract domain. Because of the abstractness, it was exceedingly difficult to formulate interview questions that would elicit the above aspects and at the same time not be too broad or to narrow. Our model is largely a synthesis of components that can be found in Friedman et al. (1997), Barndt (1994), Kahaner (1996), Meyer (1987), and Heuer (1999).

The terminology and structures adapted for this study should provide ways to characterize an instance of an intelligence operation. If various aspects of an intelligence operation can be characterized, then an intelligent agent that utilizes case-based reasoning (CBR) might reason about past operations and offer useful direction to the user about the current operation. The discussion that follows describes what someone involved in CI (competitive intelligence) might be expected to be able to do. The model focuses on the process itself and makes no commitment to any computerized system.

General phases in intelligence analysis

The process of intelligence analysis might be divided into an arbitrary number of steps. The Wisdom Builder software divides the process into four steps: requirements, collection, analysis, and report. This is probably the closest one can come to a consensus in the field. However, the results of the survey and information gathered during our literature review suggest that intelligence providers might benefit from a more structured approach that would divide the process into more cognitively manageable chunks. Two of the main problems that intelligence providers have are formulating the requirements that direct the operation and generating and evaluating multiple hypotheses.

Our model consists of seven phases that can be completed sequentially and iteratively. In addition, we identified four "mini-phases" that can be invoked during the main phases in much the same way a computer program invokes a function or method. The main phases are:

- define problem
- identify knowledge base
- target location of information
- select intelligence mode
- collect information
- analysis
- report

The mini-phases are:

- create profile
- retask the system
- counterintelligence
- archive

The main phases are completed sequentially. The retask the system mini-phase is the method of performing a loop, going back to target location of information. Figure A1 is an overview of the process.

For the purposes of this model, we identify two main actors: the intelligence provider and the intelligence user. The intelligence provider is the entity that directs and carries out the entire intelligence operation. Of course,

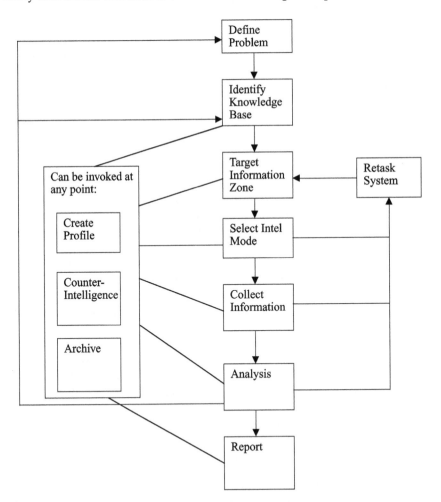

Figure A1 Overview of the competitive intelligence process

in reality, the provider may be an individual or an organization. In this model, the provider is referred to as an individual. The intelligence user is the entity that will receive the final intelligence product. The information needs and desires of this person/organization determine the direction of the operation.

Define problem

The intelligence provider receives a problem statement from the intelligence user. In general, this model distinguishes between information that the provider collects (raw information) and information that the provider has processed in some way. The problem statement is raw information. It may be in any form, from a conversation to a formal document.

The intelligence provider uses the problem statement to create a mission statement. The mission statement is the document that will guide the rest of the operation. It is made up of three parts: mission requirements, mission constraints, and user intentions. The mission requirements are the "what" of the intelligence operation. They are declarative sentences that define exactly what the intelligence provider is expected to do in order to achieve a successful intelligence operation.

The mission constraints are constraints on how the intelligence operation will be carried out and what form the final intelligence product will have. The mission constraints consist of time constraints and form constraints, but may also include other constraints which do not fit well into these subtypes. The time constraints can be either ongoing or definite. Definite time constraints can be either relative or absolute. Ongoing time constraints must be defined in terms of time between deliveries of the intelligence product to the intelligence user. These times may also be relative or absolute. Form constraints define what form the intelligence operation and final intelligence product must have in order to ensure that the user accepts them. Form constraints include the set of ethics that is appropriate to the user, the intelligence system the user trusts, whether the product should be qualitative or quantitative, the format in which the final product should be presented, etc. (Barndt, 1994).

User intentions are the "why" of the intelligence operation. User intentions describe why the intelligence user wants a certain piece of intelligence and why it is important to him/her/the organization. This information gives the intelligence provider the context to recognize information that is significant to the intelligence operation. There are many ways to determine the user intentions and form constraints. Performing a profile is one method that is discussed later.

The intelligence provider presents the intelligence user with the mission statement for his/her approval. Ideally, both the user and provider now understand one another so the intelligence user approves the mission statement. Further changes to the mission statement require interaction with the intelligence user. These changes should be rare and should be seen as profoundly affecting the course of the intelligence operation.

The purpose of defining the problem is to ensure that the intelligence provider understands the needs and desires of the intelligence user. The mission statement describes all of the criteria for a successful operation. Without a clear mission statement, the intelligence operation cannot succeed. The problem has been successfully defined when all parts of the mission statement have been completed and both the intelligence user and intelligence provider agree that the mission statement represents a common understanding.

Identify knowledge base

After the mission statement has been successfully completed and before information can be collected, the intelligence provider should explicitly identify what is known and what needs to be known. The knowledge base is the structure that contains all of this information. The knowledge base has three parts: the information inventory, the assumption inventory, and the information requirements. The information inventory is a list of succinct declarations that represent information relevant to the mission requirements. This information is regarded as factual, pending the validity of its associated source. Each declaration is associated with one or more sources. We use a stipulative definition of information for this model. Something qualifies as information only if it is associated with a source and we intend to evaluate the worth of that source. An assumption is information whose source will not be evaluated. Accordingly, the knowledge base contains an assumption inventory. Assumptions are relevant to an operation if they affect the intelligence provider's judgment. They should be explicitly identified. If they are to be used later in the analysis phase, the intelligence provider must decide whether or not the source will be evaluated. The intelligence provider uses the mission requirements, the information inventory, and the assumption inventory to create the information requirements. The information requirements are a list of core questions and their associated subquestions that target the information that the information inventory does not (or does not adequately) cover.

The goal of identifying the knowledge base is threefold: to keep track of what is known and what needs to be known, to explicitly identify the information that affects the intelligence provider's judgment, and to allow the intelligence provider to focus on answering a series of discrete, manageable questions. Identifying the knowledge base is the initial phase of setting up the knowledge base; the knowledge base will be iteratively refined and updated throughout the operation. Therefore, there are no real rules for judging the completion of this phase. The knowledge base is updated during the retask the system mini-phase. Information that is added to the knowledge base is no longer raw information. It has been considered and, perhaps, summarized by the intelligence provider. It is now part of the system.

Target location of information

Once the information requirements have been defined, the intelligence provider can begin to target the probable location of the information that will satisfy the information requirements. The term *location* needs further explanation. Friedman et al. (1997) conceptualize information as residing in one or more information zones. According to them, there are five information zones: electronically formatted (zone 1), paper formatted (zone 2), gossip (zone 3), gray (zone 4), and proprietary/secret (zone 5). The ease with which a piece of information may be retrieved, the time it will require to retrieve it, the resources required, the amount of information emission that retrieving it will produce, and the risk involved in retrieving it are determined by the zone in which a piece of information resides.

Targeting the probable location of the information that will satisfy the information requirements is done so that in the next phase, the intelligence provider will be able to make a decision about which intelligence-gathering mode (which will be explained later) to enter. The output of the identify location of information phase is an information requirements evaluation. The information requirements evaluation is the list of information requirements with each requirement labeled with the location in which it is likely to be found.

The intelligence provider uses knowledge of the nature of each of the information zones as well as past experiences to make the determination. The value of creating the information requirements evaluation is that it allows the intelligence provider to more easily compare the current information requirement from the current operation with the information requirements from previous operations in order to make a better estimation about the likely location of the required information. This phase has been successfully completed when all of the information requirements have been mapped to one or more information zones.

Select intelligence-gathering mode

At this point, the intelligence provider has an idea of what information is needed and what kind of resources will be consumed in order to obtain this information, and must now decide which information will be pursued. This decision is determined by the relationship between the available resources (time, money, equipment, and personnel), the estimated costs (estimated from information zone characterizations), and the value of the required information to the intelligence user.

Friedman et al. (1997) distinguish three modes of intelligence gathering: passive, semi-active, and active (explained further in the next section). In general, the intelligence provider should begin collection in the passive mode and only move to the semi-active and active modes if absolutely necessary. The information that is chosen to be pursued determines the intelligence-gathering mode. This decision should be made explicitly, because each mode of intelligence gathering requires its own preparation. The modes are differentiated by various factors, many of which are inherent in the information zone with which they are associated. However, one of the most important factors that differentiates the modes is the amount of information that is emitted by the intelligence provider in the course of collecting information.

The goal of this phase is to encourage the intelligence provider to methodically weigh the consequences, costs, and benefits of entering a particular mode of intelligence gathering. The issue of emitting information that can be detected by others is important, because such an emission may destroy the competitive advantage that the intelligence product would have to the user and hence cause the operation to fail.

Collect information

After weighing the alternatives, the intelligence provider begins to pursue the information that was targeted in the previous phase. In general, the intelligence provider should begin by mining information that is internal to the organization and then move to external sources (Friedman et al., 1997). The reasons for doing this are as follows. An organization already has a mechanism in place for gathering tons of information every day. Much of that information is gathered by people who already share the context of the organization, and they are likely to be in a position to recognize significant information when they see it. Outside sources are expensive, do not share the context of the organization, and, therefore, may be less productive.

The passive mode is the mode that should always be entered first. In this mode, the intelligence provider gathers information from zones 1 and 2. All of this information exists in the public domain and is free to anyone. Collecting this information involves very little time, skill, expense, or social interaction. Since it is freely accessible, collecting this information does not leave traces that could be detected by a competitor who is attempting to discover one's intentions. A rule of thumb for determining when zones 1 and 2 have been exhausted is when the bibliographies of newly collected pieces of information contain references to the things that you have already gathered (Friedman et al., 1997). When the first two information zones appear to be exhausted, the system should be retasked.

The semi-active mode should be entered only if zones 1 and 2 have been exhausted, there are still unsatisfied information requirements, and the value of the required information to the user exceeds the costs and risks associated with retrieving it. This mode involves collecting information from zones 3 and 4 and demands social interaction, so appropriate measures such as performing a personality profile, choosing a persona, identifying the appropriate contact method, and tracing social networks are required. For these reasons, the semi-active mode is slower, more expensive, and more difficult than the passive mode. The extra measures required by this mode also increase the amount of information emitted by the intelligence provider's actions.

For each instance of social interaction, a contact sheet should be created. The contact sheet consists of the name of the source, the task associated with this instance of interaction, the method of contact, the persona used, the results of the interaction (including the information retrieved), and further questions that resulted from this interaction. A rule of thumb for determining when zones 3 and 4 have been exhausted is when sources indicate other sources that have already been examined.

The active mode involves gathering information from zone 5, the proprietary/secret zone. It requires the presence of the intelligence provider (or some agent, either mechanical or human) to detect an information emission (Friedman et al., 1997) from a target. This may require the organization of covert operations that includes agents and subagents to penetrate a target organization.

Retasking the system

Retasking the system is the mechanism for performing a loop in the larger intelligence analysis process. It can be invoked at any point in the process, but it will most often occur after the completion of an iteration of the collection phase (in one mode or another). In this phase, the intelligence provider deliberately moves information, which has been collected and intermediately stored, into the knowledge base. This means that each piece of collected information has been processed or summarized in some way by the intelligence provider and it will now officially become part of the system. Information should be moved into the knowledge base if it affects the intelligence provider's judgment.

Once the information inventory and the assumption inventory have been updated, they should be mapped against the information requirements. The information requirements that are not covered by the two inventories are the gaps in the knowledge base. New information requirements should be added to these to comprise the updated information requirements list.

Analysis

In this phase, the intelligence provider generates possible solutions (hypotheses) for each of the mission requirements that are contained in the mission statement. Multiple hypotheses should be generated for each mission requirement. These hypotheses are then evaluated with regard to the evidence that is associated with them. A piece of information from the knowledge base becomes evidence when it either supports or undermines a hypothesis.

According to Heuer (1999), the steps for evaluating multiple hypotheses are:

- Create a matrix of solutions and evidence.
- Create for/against lists for matching the information with hypotheses.
- Remove irrelevant information.

- Evaluate the diagnosticity of evidence: for each piece of evidence, the intelligence provider counts the number of hypotheses it supports. The fewer hypotheses a piece of evidence supports, the more diagnostic it is.
- Remove evidence with no diagnostic value.
- Assess the likelihood of each hypothesis.
- Determine sensitivity (the sensitivity reference points to the pieces of evidence upon which a hypothesis depends).
- Identify key events.

The goal of this process is to leverage all of the information that the intelligence provider has collected and reduce the cognitive burden inherent in evaluating multiple hypotheses. After all of the hypotheses have been evaluated, the intelligence provider must decide on a recommendation. A recommendation is a concise declaration of what the intelligence provider believes is the solution to the mission requirements. Its purpose is to support the intelligence user's ability to make a decision rather than just supplying him/her with more information.

Report

Once the collection and analysis phases are complete, the intelligence provider's findings must be put in a form that provides value to the intelligence user. The intelligence product consists of the evidence, hypothesis evaluations, and recommendations. The report can also include the subphases: create profile (mini-analysis phase that is used to acquire certain kinds of information); counterintelligence (intelligence provider identifies which pieces of information represent a competitive advantage to the intelligence user); and archive (consists of storing all of the information that makes up a certain intelligence operation according to the terms or features that characterize the operation).

Conclusions and future directions

A commitment to some model of an intelligence analysis process is necessary to provide a computerized agent with knowledge about what steps make up the "correct" intelligence analysis process. Ideally, the intelligence provider can customize this model, to a certain extent, in order to meet his/her needs. A tool called POINT (Problem Organization INtelligence Tool) has already been developed by the authors, using Visual Basic and MS-Access, to encode the model presented here. This tool and the encoded model are now being tested and evaluated by analysts at the Federal Bureau of Investigation (FBI).

The most challenging issue concerns defining the stages of the intelligence analysis process where the agent can provide valuable assistance to the intelligence provider. The terms and artifacts introduced by this model provide a means to characterize instances of intelligence operations. For example, as the intelligence provider creates the information requirements, an agent could compare terms contained in this artifact with terms contained in the information requirements evaluations of previous operations. By then identifying the information zones that were necessary to access as well as the sources that yielded relevant information, the agent could suggest useful advice about which intelligence mode to enter or the potential cost of acquiring certain information. The agent could also take the initiative to retrieve information from sources that have been useful in the past, and, hence, save the intelligence providers some time.

Another possibility is that, during the analysis phase, an agent could notify the user when a certain likely hypothesis relies on an important piece of evidence that is either an assumption or that comes from a source that has been unreliable in the past. We are exploring text mining and case-based reasoning further to possibly help in those areas. It is hoped that more opportunities such as those mentioned above will become apparent as we continue to consider the intelligence analysis process as a collaboration between an intelligence provider, an agent, and the system (the model).

References

Barndt, W. (1994), *User Directed Competitive Intelligence: Closing the Gap Between Supply and Demand*, Quorum Books, Westport, Connecticut.

Friedman, G. et al. (1997), *The Intelligence Edge: How to Profit in the Information Age*, Crown Publishers, New York.

Heuer, R. (1999), Psychology of Intelligence Analysis, Center for the Study of Intelligence, Central Intelligence Agency, www.odci.gov/csi/books/19104/index.html.

Intelligent User Interfaces Conference Proceedings (1999), Association for Computing Machinery, New Orleans.

Kahaner, L. (1996), *Competitive Intelligence*, Simon & Schuster, New York.

Meyer, H. (1987), *Real World Intelligence: Organized Information for Executives*, Weidenfeld & Nicolson, New York.

Schreiber, G., H. Akkermans, A. Anjewierden, R. de Hoog, N. Shadbolt, W. van de Velde, and B. Wielinga (2000), *Knowledge Engineering and Management: The CommonKADS Methodology*, MIT Press, Cambridge, MA.

appendix D

Planning and scheduling in the era of satellite constellation missions: a look ahead*

Jay Liebowitz

Scope of the study

As NASA Goddard looks ahead in the next few years, they will see scientific satellite missions that are quite different from the missions of today. One major difference is that a group of satellites, from 3 to 100, will form a constellation and will share and distribute critical activities, data, and functions among themselves. This varies from today's traditional single satellite mission. One of the key functions that will be essential as part of this satellite constellation concept is "planning and scheduling."

The focus of this study is to examine planning and scheduling in terms of a satellite constellation mission. As such, there are several key questions for this study:

- How will satellite constellation missions be different from today's missions?
- What implications will this have for planning and scheduling functions?
- How can the planning and scheduling functions be distributed among the satellites and perform replanning/rescheduling in an optimal way?
- What technologies, tools, and techniques can aid in this process?

Each of these questions will be addressed in turn.

---

* The author would like to thank Henry Murray for sponsoring this study and for his vision for looking ahead to the era of satellite constellation missions.

How will satellite constellation missions be different from today's missions?

In the current environment at NASA Goddard, the satellite ground control works as follows (Brann, Thurman, and Mitchell, 1996):

> Human operators periodically establish a communications link (known as a 'pass') with an unmanned scientific satellite to transmit scientific data from the satellite's instruments to the ground, assess the health and safety of the satellite, uplink commands which enable it to function until the next period of communication, and monitor the communications links between the control room and the satellite. Data from the satellite are transmitted through a ground station and arrive at a control room in which operators monitor and control the satellite via a network of computer resources.

Spacecraft ground systems at Goddard are used to command the spacecraft and to capture the telemetry and data sent from the spacecraft (Stoffel and McLean, 1997). Prepass activities include planning and scheduling of the communications resources and of the spacecraft activities to be executed by the next upload. Pass activities include initiating pass scripts and sending commands, monitoring telemetry, snapping specific telemetry pages and printing them, and responding to critical events. Postpass activities include initiating the closing of the history and attitude data files, transferring engineering data to a trend analysis workstation, checking the spacecraft status buffer for errors, logging acquisition of signal and loss of signal times, and describing all tasks performed during the pass (Stoffel and McLean, 1997).

As we look toward the future era of satellite constellation missions, functions similar to those described above will need to be conducted. However, there will be a greater need for coordinating the functions across multiple satellites in a constellation. Automation will play an increasing role in accomplishing the planning and scheduling functions in a satellite constellation. However, Brann, Thurman, and Mitchell (1996) suggest that operator–automation interaction is critical to the ultimate success of sophisticated autonomous control systems as we move toward a lights-out automation approach. For effective overall system operation, the design of the automation must support operator functions such as inspection, prediction, repair, and maintenance.

According to Peter Hughes at NASA Goddard, a major step for NASA in remote sensing and space exploration is in the development of teams of satellites — also called constellations or formations of satellites (Chweh, 1998). Hughes believes that such constellations of missions will require increasing spacecraft autonomy to monitor and control the spacecraft's health and activities.

In today's telecommunications environment, satellite constellations already exist. For example, the November 29, 1999 Cambridge Telecom Report noted that the twelfth successful launch completes Globalstar's operational 48-satellite constellation. The Iridium Low Earth Orbit Satellite System is another example of a constellation of 66 satellites for communications (Pratt et al., 1999). In addition to Globalstar and Iridium, Orbcomm, Odyssey, and INMARSAT-P are all satellite constellations. In the military environment, the impact of emerging commercial satellite systems on joint operations for command and control in 2010 has been discussed, for example, by Viginia Ashpole and Theresa Clark at the Air University.

Looking toward NASA's new millennium program and NASA Enterprise, in-space Internet technologies for NASA Enterprises' revolutionary communications concepts are being worked on at such places as NASA Glenn Research Center/Space Communications (Mudry, 2000). NASA Glenn Research Center is developing advanced network technologies, techniques, and methodologies for communications within constellations of spacecraft and for communications and command functions sent between the constellation and directly to ground users.

Additionally, the new millennium autonomy integrated product development team leads the development and validation of autonomy technologies needed to fulfill NASA's vision of 21st century spacecraft and ground operations capabilities and functions (http://www.jpl.nasa.gov/tec/ipft/auto.html). With respect to planning and scheduling issues, this team is looking at planning and replanning, contingency planning, onboard sequencing validation, and conflict detection and resolution.

According to Yunck et al. (1995), the full promise of spaceborne global positioning system (GPS) science can be realized only with dedicated constellations of orbiting GPS receivers specifically designed for scientific use. Initial proposals have been for a pilot constellation of a dozen or so microsats launched at once into a single orbit plane. In the future, we will see a large constellation of hundreds of tiny satellites, each with a mass of less than 1 kg, enveloping the earth in multiple orbit planes. Scientists at JPL (Jet Propulsion Laboratory) estimate that at least 100 satellites will be needed to usefully enhance weather prediction (Yunck et al., 1995). The concept for a pilot constellation of spaceborne GPS receivers for earth science could be having twelve GPS-equipped microsatellites launched at once into a single orbit plane for a science mission of global climate and weather modeling, ionospheric imaging, and long-wavelength gravity recovery. The technology could be precision GPS-on-a-chip, auto-navigation, auto-spacecraft, and auto-operations.

A trend toward multiple-spacecraft missions is developing. Cluster II has four spacecraft for multipoint magnetosphere plasma measurements. One proposed interferometer mission would have 18 spacecraft flying in formation in order to detect earth-sized planets orbiting other stars. Another proposed mission involves 5 to 500 spacecraft orbiting Earth to measure global phenomena within the magnetosphere (Barrett, 1999).

A distributed architecture for handling planning and scheduling in satellite constellations has been proposed by a number of researchers (e.g., desJardins et al., 1999; desJardins, 2000). An agent should engage in distributed planning when planning knowledge or responsibility is distributed among agents or when the execution capabilities that must be employed to successfully achieve objectives are inherently distributed. However, a major challenge with the distributed planning approach is the synchronization problem. The synchronization problem in distributed planning becomes even greater when the distributed agents' plans are being executed concurrently, and better models of the overall planning and execution process are needed. Additionally, methods for distributed plan repair, as well as distributed plan execution, need to be created (desJardins et al., 1999).

What implications will this have for planning and scheduling functions?

Liebowitz (1999) interviewed some key NASA Goddard personnel and asked the following question: "How do you envision the ground system (GS) and on-board systems working in 10 years?" The responses included:

- Put GS functions onboard the spacecraft (onboard processing).
- All spacecraft health and safety is performed by the spacecraft itself.
- The spacecraft is preprogrammed (scripted) to perform the tasks defined for its mission life.
- The spacecraft will ping the ground when it needs to dump data or needs help.
- The spacecraft will use plug and play architectures — the only unique pieces being the instruments and the science software.
- The spacecraft will have its own site on the Web to check status.
- There will basically be no ground system.
- Move data with a low cost Internet connection — get data on-demand.
- There will be no flight operations controllers, just instrumentors operating their systems.

According to Cooke and Hamilton's article (2000) looking at new directions at NASA Ames Research Center, automated reasoning for autonomous systems will play a major role in the satellite constellation mission era. Onboard planners will reduce mission and operational costs by greatly decreasing the amount of human effort currently required to plan, schedule, and execute detailed sequences of vehicle commands.

Remote Agent is an example of an autonomous onboard planner and is the first artificial intelligence (AI) software in history to command a spacecraft, Deep Space One. Remote Agent consists of three components: the planner/scheduler (PS), smart executive (exec), and mode identification and reconfiguration (MIR). When requested to do so, PS generates a plan that exec breaks down into tokens that are sent to the flight software to execute. MIR listens in and builds a model of what the spacecraft's state should be and contingency plans in case the program fails to execute as expected. Originally, Remote Agent was written in Lisp but is now being rewritten in C++.

With the satellite constellation mission concept, planning and scheduling functions will perhaps be conducted in a distributed manner. Various research on cooperative/negotiated distributed planning and distributed continual planning is being conducted at NASA, industry, and various research institutions. DesJardins et al. (1999) identify various research issues being addressed in these areas. Cooperative distributed planning places the problem of forming a competent (sometimes even optimal) plan as the ultimate objective. It is typically carried out by agents that have been endowed with shared objectives and representations by a single designer or team of designers. The key components are:

- plan representation and generation (e.g., abstraction-based plan representation, partial global planning)
- task allocation (load balancing)
- communication
- coordination

Negotiated distributed planning places the emphasis on having an agent provide enough information to others to convince them to accommodate their preferences. Key elements include:

- collaboration
- negotiation (negotiation strategies, voting schemes, contract nets)

Future research directions relating to distributed planning include:

- reasoning and negotiation techniques
- open-ended planning, in which plans can continually be refined to varying levels of abstraction as the planning horizon is extended
- need for better models of the overall planning and execution process

Other interesting research issues in these areas include finding the right balance between a plan's flexibility and robustness and the speed of the computational substrate needed to construct and execute the plan (Chien et al., 1998). Another issue is how to integrate planning and execution to allow for responsiveness to run-time variations. Other research areas involve representing preferences, generating optimizing planners, and gaining a better understanding of the relationship between spacecraft operations planning and other types of scheduling.

According to the LOGOS (lights-out ground operations system) work (LOGOS PSWG Working Group, 1998) conducted at NASA Goddard, the planning and scheduling working group determined the following list of preliminary requirements for a planning engine and executor features in an automated, lights-out scenario:

- able to generate partial-order plans based on submitted goals
- able to generate partial-order plans based on incomplete plans
- able to generate plans using actions as well as sub-plans (hierarchical task network planning)
- able to execute partial-order plans and monitor the execution of actions for success and failure (This should be a separate component/module.)
- able to work with goals, preconditions, and effects described with first-order predicate logic or its equivalent
- able to reason about actions which have measures or resources, especially time
- able to use different search strategies (depth-first, breadth-first, A*, depth-first iterative deepening, etc.)
- deterministic if desired, i.e., returns the same solution with the same starting conditions

Other features, such as disjunctive and universally quantified effects and preconditions, conditional planning, and learning, are useful but may not be necessary for these projects.

The LOGOS planning and scheduling working group also conducted a survey of existing planning and scheduling systems: MOPSS, ROSE, UCPOP, Graphplan, HBSS, GUESS, GenH, NM-DS1, APS, and GENIE. These systems were partially evaluated using the following planning and scheduling tool features:

Planning tool features:

- generation
- execution/monitoring
- replanning
- partial order
- temporal reasoning
- resource reasoning
- alternative search strategies
- hierarchical/approximation

Scheduling tool features:

- dynamic rescheduling
- temporal constraints

- periodic constraints
- consumable resource reasoning
- durable resource reasoning
- event generation
- event reasoning
- conditional scheduling

The Canadian Air Force also identified a set of major planning/scheduling tool features for their set of problems, which compares favorably with the NASA-generated requirements:

- replanning capability
- adaptive planning
- dynamic, time-constrained, and uncertain environment
- long-, medium-, and short-term planning
- intelligent mission planning based on criteria
- AI functionality
- schedule optimization
- mission pre-emption
- automatic rescheduling
- what-if analysis
- missions merging
- airframes and aircrews scheduling by events
- flight schedule and flying programs support
- resources and equipment assignment
- flight time compilation tables
- alternative plans storage

How can planning and scheduling functions be distributed among the satellites?

In most practical problems, planning and scheduling are accomplished by multiple agents as opposed to a single agent. A key challenge is to enable all agents to generate their plans with some (acceptable to all) degree of autonomy and independence, and yet provide a mechanism for them to harmonize the resulting plan and to optimize it toward the greater good. The TRAC2ES system, developed by Logica Carnegie Group, uses a "deconflicting agent" to collect plans from all regional planners, generate a set of corrections and then recommend them to the regional planners. In the JMISS system, also by Logica Carnegie Group, agents are provided with the protocol that enables them to continuously exchange, while developing their individual plans, their current demands for various resources.

According to the 1998 AAAI Fall Symposium on Distributed Continual Planning, distributed planning means that the planning activity is distributed across multiple agents (Weng and Ren, 1997), processes, and/or sites.

Continual planning (sometimes called continuous planning) means that the planning process is an ongoing, dynamic process. Topics that are relevant to these areas are:

- Representations and models for interagent cooperation and communication, including machine–machine, machine–human (mixed initiative), and human–human (collaborative) interactions
- Continuous planning — i.e., managing an environment in which planning is interleaved with execution and other related activities (scheduling, simulation, evaluation), objectives are dynamically changing, and new information about the state of the world arrives continuously and asynchronously
- Technologies for splitting a planning problem into subproblems, coordinating the activities of subplanners (e.g., sharing constraints), resolving conflicts among subplans, and merging the resulting subplans
- Architectures and infrastructure support for distributed planning systems (e.g., hierarchical models of distribution, workflow and process management, plan servers, and repositories for shared knowledge)
- Meta-level control strategies for allocating an agent's computational resources to support multiple tasks

Liebowitz and Potter (1995) surveyed the expert scheduling system literature and discovered that there were three main classes of scheduling approaches used in expert scheduling systems:

1. Optimization, including enumerative algorithms and mathematical programming methods
2. Heuristic, which can be broken down further into:
   - Heuristic algorithms:
     - Neural network algorithms
     - Greedy algorithms
     - Intelligent pertubation algorithms
     - Simulated annealing algorithms
     - Genetic algorithms
   - Heuristic scheduling methods:
     - Resource-based scheduling
     - Order-based scheduling
     - Hierarchical decomposition
     - Incremental scheduling
     - Reactive scheduling
     - Conflict resolution strategies
     - Activity-based scheduling
     - Requirements-driven scheduling
     - Blackboard-based scheduling
     - External expansion with backtracking

- Repeat–expand cycles
- Internal expansion with backtracking
- Look-ahead scheduling
- Delayed evaluation strategy
- Constraint-based representation

3. Hybrid, which includes AI + simulation-based scheduling and optimization/operations research + AI scheduling.

Tsang's (1995) comparative study of scheduling techniques identified the classic scheduling techniques developed in operations research as linear programming, branch-and-bound, and tabu search. Better known scheduling techniques in AI are hill climbing, simulated annealing, genetic algorithms, and expert systems. Some key questions to ask when choosing which scheduling techniques to use are (Tsang, 1995):

- Does the problem require any solution which satisfies all the constraints, or does it require some sort of optimization (i.e., satisfiability vs. optimization)?
- If optimization is required, is the function to be optimized linear?
- If optimization is required, are near-optimal solutions acceptable?
- How much time does the program have for finding solutions?
- What sort of domain-specific knowledge is available?

In general, predictability of schedule optimality is orthogonal to schedule optimality, and they must be traded off against one another to application-specific requirements. For example, a deterministic soft real-time scheduling algorithm may not always (or even ever) produce the optimal schedule, but it may produce acceptably suboptimal schedules (http://www.realtime-os.com).

At the March 2000 NASA Planning and Scheduling Workshop, the main discussion issues were (http://ic.arc.nasa.gov/ic/psworkshop): responsiveness, validation (how to guarantee that a plan generated in a previously untested situation will indeed operate the system correctly and safely), mixed-initiative autonomy, and mission acceptance.

Some researchers are using intelligent retrieval agents, called mediators, that execute prespecified guidelines of how to combine and integrate information from different sources (Davydov, 1999). They have applied a federation of mediators and reasoning agents, called facilitators, that are based on deductive databases implemented in Java and can reason with metadata sources.

Automated planning and scheduling technology can hopefully lead to reduced costs; increased responsiveness, interactivity, and productivity; and simplified self-monitoring. Three of NASA's planning systems projects for deploying automated planning technology for spacecraft command are: Dcaps (data-chaser automated planning system), Deep Space One, and Aspen. Dcaps, developed jointly by JPL and the University of Colorado, showed that basic planning technologies — such as initial-schedule generation and basic heuristic iterative repair — can be of considerable value in operations. NASA's new millennium program's first mission was Deep

Space One, which contained the Remote Agent, in which a PS planning system was included. The Aspen planning system is being used for fully automated, lights-out commanding (Chien et al., 1998).

Zoch and Hull (1991) believe that their NASA Goddard-sponsored SAIL (scheduling applications interface language) could be used in a distributed planning and scheduling network. Barrett (1999) and the planning and scheduling group of the artificial intelligence group at JPL have been examining autonomy architectures (Hexmoor and Desiano, 1999) for a constellation of spacecraft. Different architectures have been proposed. One approach is the master/slave coordination whereby the constellation is treated as a single spacecraft (as the master) with virtually connected slaves. In this approach, one spacecraft directly controls the others as if they were connected. The controlling master spacecraft performs all autonomy reasoning while the slaves only transmit sensor values to the master and forward control signals received from the master to their appropriate local devices (Barrett, 1999).

Another approach is to have a team of spacecraft contain a leader and one or more followers that jointly intend to accomplish some task by executing a team activity. In the master approach described above, the master knows everything about the constellation, whereas the team leader only knows a subset of everything (Barrett, 1999). This approach uses collaborative planning which involves distributing the plan across the constellation and letting each spacecraft detect and repair problems. The main question here is how to keep the plan consistent across the constellation while all spacecraft are updating it. The main objective is minimizing communications overhead while planning (Barrett, 1999). To do this, one method would be to fragment the plan and distribute the fragments. Since the fragments are disjointed, their union would be consistent. An alternative approach is to give every spacecraft a copy of the plan and have them maintain consistency by broadcasting changes as they make them. However, again, the communications overhead could be a major problem since the spacecraft would spend most of their time responding to each other's updates (Barrett, 1999).

Architectures can co-exist within a constellation. For example, a constellation could have three classes of spacecraft — leaders, followers, and slaves. Leaders have the ability to plan and collaborate. Followers can only execute plans and watch out for each other. Both leaders and followers can have virtually attached slave spacecraft (Barrett, 1999).

What technologies, tools, and techniques can aid in this process?

According to Liebowitz's interviews (1999), one of the future goals is to re-engineer planning and scheduling. Technologies to explore in the near future include using heuristic, constraint-based programming, neural network, and genetic algorithm techniques to improve planning and scheduling. Design tools using interactive simulation techniques for mission planners to run what-if scenarios should be developed. Longer-term technologies to consider include chip manufacturing, which may allow scheduling algorithms to be hard coded onto a chip (similar to fuzzy logic chips).

NASA Goddard's GenSAA modular expert systems can be built for a discrete subsystem or function and can execute concurrently and share key conclusions with one another using a publish-and-subscribe model of communication. That way, system level expert systems can simply monitor the conclusions made by the subsystem level expert systems. The GenSAA data server would be needed in this case as a central repository in which GenSAA expert systems can publish information and from which other subscribing GenSAA expert systems can receive the information when published.

At the Distributed Intelligent Systems Laboratory at the University of Massachusetts–Amherst, work has been conducted (dis.cs.umass.edu/research/arm.html) in negotiation among knowledge-based scheduling agents. Researchers there developed the DENEGOT architecture which serves as a general approach for guiding the distributed search of multiple agents through negotiation to arrive at a satisfying solution to a possibly overconstrained problem. They have also applied these techniques in a distributed airline resource manager application. The primary difficulty in distributed scheduling is that no scheduling agent possesses a global view of the problem space. To alleviate this problem, meta-level communication and on-line negotiation between the agents are being researched.

Stottler Henke Associates Inc. (SHAI), under the leadership of Rob Richards, is working on an SBIR phase I grant for Code 583 examining the issue of planning and scheduling for spacecraft coordination during constellation missions. Their proposed architecture will support the creation and maintenance of adaptive, hierarchical, organizational structures to the collaborative planning process among satellites (http://www.shai.com/projects/satellite4.htm). It will also use a sophisticated knowledge base to store information about satellites' capabilities and commitments in order to allocate tasks to the satellites best suited to perform them. The primary AI techniques used in this project are automated planning and scheduling, constraint satisfaction, and knowledge-based representations.

In Chien et al.'s work (1999) at JPL, an on-board planner/scheduler is considered to be a key component of a highly autonomous system for spacecraft operations. They found that automation of the commanding process can reduce mission operations costs by as much as 60 to 80%, as compared with manual generation of sequences. To achieve a high level of responsiveness in a dynamic planning situation, Chien et al. utilize a continuous planning approach and have implemented a system called CASPER (continuous activity scheduling planning execution and replanning). The basic algorithm for this continuous planning approach is:

Initialize P to the null plan
Initialize G to the null goal set
Initialize S to the current state

Given a current plan P and a current goal set G:

1. Update G to reflect new goals or goals that are no longer needed.
2. Update S to the revised current state.
3. Compute conflicts on P, G, S.
4. Apply conflict resolution planning methods to P (within resource bounds).
5. Release relevant short-term activities in P to RTS (real-time system) for execution.
6. Go to step 1.

Two of the major commercial constraint-based scheduling tools are Cosytec's CHIP constraint programming system and ILOG's constraint-based solver/schedule/views/optimization software. The ILOG optimization software has been used for resource allocation for the telecommunication satellite constellations (Freuder and Wallace, 2000).

There has been software designed especially for the visualization and analysis of satellite constellations. One example is SaVi, developed by researchers at the University of Minnesota. SaVi provides a three-dimensional visualization of satellites in orbit around the Earth, a display of satellites' footprints on the Earth's surface, the computation of the fraction of the Earth's surface covered by the constellation, and simulations of a view of the satellites overhead from an arbitrary point on the Earth's surface. It runs only in a Unix/X11 environment (including Linux on a PC). SaVi does not run on a Windows operating system. It is freely available by e-mailing savi@geom.umn.edu. COLLAGE is a planning tool for data analysis tasks being used in the satellite domain and is available from the same source.

The airline industry has been using planning and scheduling tools for their applications. OpsSolver, by Caleb Technologies Corp. (www.calebtech.com/OpsSolver.htm), is one such tool. It uses state-of-the art optimization techniques to recover flight schedules during irregular operations. Utilized by Continental Airlines, OpsSolver optimizes schedule recovery in response to disruptions. ZPR, a German company, develops optimization tools for the airline industry (www.zpr.uni-koeln.de). Their work focuses on developing planning tools to improve operations at airlines, namely fleet assignment problems and crew scheduling problems. Their work has been applied at Lufthansa.

According to James Martin Strategies (JMS North America Inc., 1995), the resource planning group at USAir is responsible for scheduling flights up to 18 months in the future. Their parameters include aircraft, flight crews, maintenance staff, and facilities, as well as gate allocation at both the departure and arrival airports. The schedules are generally finalized 45 days before implementation. The Schedule Development Environment (SDE) at USAir uses a scheduling system from SSI, an MIT spin-off, developed specifically for the airline industry. The package has been customized to USAir's operations. The system has been described by the resource planning staff as extremely user friendly, requires only a short learning curve, and allows several staff members to work on a schedule at the same time. Schedule conflicts are indicated immediately, but overrides are allowed. The level of constraints is determined by the user.

Montreal-based Ad Opt Technologies' Preferential Bidding System (PBS) schedules flight crews for TWA, Delta, and Air Canada. It costs between $500,000 to $3 million, depending on the different options and size of the application. SARANI (//konark.ncst.ernet.in/~kbcs/sarani.html) is another airline flight scheduling system developed by NCST. It is an intelligent airline flight scheduling system used by Air India. American Airlines set the industry standard for airline reservation systems with their product, named SABRE. The SABRE Group (www.sabre.com) is owned by American Airlines and consists of a consulting group that develops planning and scheduling applications, as well as a number of other airline-related tasks.

In Tsang's (1995) comparative analysis of scheduling techniques, the following conditions for applying specific scheduling techniques were determined:

- Linear programming
    - Used for optimization with linear functions
    - Intractable
    - Specification of problem by a normally conjunctive set of equalities
- Branch-and-bound
    - Used for optimization
    - Intractable
    - Requires heuristic for pruning
    - Ordering of branches important
- Constraint satisfaction
    - Most existing algorithms used for finding single or all solutions satisfying constraints
    - Both complete and incomplete algorithms available
    - Large number of available algorithms particularly useful when problem involves nontrivial amount of constraints
- Hill climbing/simulated annealing/tabu search
    - Useful for both constraint satisfaction and optimization when near-optimal solutions are acceptable
    - Flexible in computation time, which makes them widely useful
    - Hill climbing possibly trapped in local optima
    - Simulated annealing and tabu search attempt to escape from local optima
    - Hill climbing requires a neighborhood function which is crucial to its effectiveness
    - Simulated annealing's neighborhood function is crucial to its effectiveness, and cooling schedule could be important

- Tabu search's effectiveness depends mainly on strategy on tabu-list manipulation
- Genetic algorithms
  - Useful for satisfiability problems or for finding near-optimal solutions
  - Good potential for parallel implementation which may suit real-time applications
  - Set up and network updating mechanism crucial to its effectiveness
  - Specialized network potentially expensive to build
- Expert systems
  - Wide range of applicability (can be tailor-made to meet the requirements including time and optimality)
  - Power comes from domain-specific knowledge
  - Expert knowledge elicitation is important and may be difficult
  - Conflict resolution may be nontrivial

Much of the work in the planning community has been focused on disjunctive planners. Disjunctive planners retain the current plan set without splitting its components into different search branches (Wilkens and desJardins, 1999). Concern exists that disjunctive planners may not be practical for real-world planning applications, and hierarchical task network planners and knowledge-based planners may be an alternative approach to explore. An emphasis on developing richer domain knowledge models is necessary in order to make planning tools useful in complex problems.

According to Wilkens and desJardins (1999), an ideal system would have to:

- Exhibit creativity, devising new actions that can solve a problem or shorten a plan
- Use analogy to transfer solutions from other problems
- Effectively interact with humans to use their knowledge in decisions
- Behave intelligently in the face of conflicting or incomplete information

Wilkens and desJardins (1999) believe that these capabilities will require more knowledge, including background knowledge of other domains and of how the world works. However, not all interesting problems require the above characteristics. For example, the Deep Space One planner used purely propositional representation for encoding the configuration planning and execution system.

Wilkens (1998) describes a multi-agent planning architecture, developed under an ARPA/Rome Laboratory planning initiative grant. MPA uses a range of generic planning agents that provide specific services in response to a range of requests. These agents are capable of reporting incremental progress, providing partial plans, and continually responding to new constraints, conditions, and suggestions. Meta-planning agents (planning agents that control other planning agents) are used to coordinate the activities of these planning agents. This architecture has been successfully demonstrated in air campaign planning.

Looking ahead at intelligent scheduling tools and technologies, the following research is needed:

- Benchmarking intelligent scheduling systems and performance analysis
- Establishing a methodology for building intelligent scheduling systems that clarifies the scope of their applicability
- Developing general or generic intelligent scheduling systems/tools
- Developing problem description vocabulary, classification scheme, and map of the problem space
- Constructing a mapping between problem leaf nodes and solutions leaf nodes (i.e., mapping problem requirements to scheduling approaches for use in expert systems)
- Realizing that the AI part of the scheduling system may comprise very little in terms of the overall scheduling system development
- Researching the relationship between and integration of planning and scheduling techniques (unified framework)
- Applying machine learning techniques in scheduling

Recommendations

Based upon the literature review, analysis, and careful consideration of the NASA Goddard satellite constellation mission applications, the master–slave architecture proposed by Barrett (1999), is recommended, whereby the controlling master spacecraft performs all autonomy reasoning while the slaves only transmit sensor values to the master and forward control signals received from the master to their appropriate local devices. This may be the easiest way to adapt autonomous spacecraft research to controlling constellations by treating the constellation as a single spacecraft.

A deconflicting agent approach, as proposed by Logica Carnegie Group in their TRANS2ES application, should also be applied for resolving conflicts that will occur in the planning and scheduling activities of the satellite constellation. This "deconflicting" agent should use constraint satisfaction techniques, with which a large number of algorithms and tools are available (e.g., Ilog's solver/schedule constraint-based satisfaction optimization tool, which has been used already for resource allocation of telecommunications satellite constellations), and these approaches are particularly useful when the problem involves a nontrivial amount of constraints, as in the NASA Goddard applications. Based upon the GUESS work (Liebowitz and Potter, 1995), there would typically be about 6000 weekly events and over 30,000 constraints for weekly scheduling of experimenters' usage of a NASA Goddard-supported spacecraft. Chien et al.'s basic algorithm (1999) for a continuous planning approach could also apply as:

- Initialize P to the null plan, G to the null goal set, and S to the current state.
- Given a current plan P and a current goal set G:

  1. Update G to reflect new goals or goals that are no longer needed.
  2. Update S to the revised current state.
  3. Compute conflicts on P, G, S.
  4. Apply conflict resolution planning methods to P (within resource bounds).
  5. Release relevant short-term activities in P to a real-time system for execution.
  6. Go to step 1.

Iterative repair methods for resolving conflicts (for step 4 above) and performing scheduling should be strongly considered for use in the satellite constellation era. Already, iterative repair algorithms have been used for a number of NASA planning and scheduling applications such as (Chien et al., 1999): the GERRY/GPSS system uses iterative repair with a global evaluation function and simulated annealing to schedule space shuttle ground processing activities; the Operations Mission Planner system uses iterative repair; the Space Telescope scheduling systems use the minimum-conflicts heuristic for scheduling space telescope usage; the OPIS system using iterative repair techniques; the Remote Agent Experiment Planner uses iterative repair and local search techniques. Hierarchical task networks should also be further explored for possible use in the satellite constellation planning and scheduling applications.

An interesting approach could be a multi-agent approach within the master controlling spacecraft to conduct planning for a satellite constellation. Specifically, there may be four key agents: (1) agenda developer, (2) agenda/plan tracker, (3) planner, and (4) strategic planner. The agenda developer would deal with "today's plans." The agenda/plan tracker would track the plan. The planner would contain routines to determine plans and replan if new activities occur. The strategic planner would check to see how the activities fit into an overall global plan to reduce the likelihood of local (vs. global) optimization.

This approach may not get the optimum plan, because most real-world scheduling applications (like the satellite constellation scheduling application) are NP-hard problems which may be intractable due to the many events, resources, and constraints that need to be scheduled and rescheduled. Thus, heuristics are used (e.g., iterative repair heuristics) to limit the chance of the combinatorial explosion of alternatives, and the plan/schedule should be a very satisfying solution.

The issue of what should be handled by the ground software and the flight/onboard software is an interesting one. The recommendation of this author is to let the onboard scheduling system in the master spacecraft perform the automatic planning and scheduling (and rescheduling) functions for the satellite constellation, with possible overrides by the human operators on the ground. In extremely difficult situations, the human operators may be paged (as in the LOGOS paging agent) to provide suggestions for repairing and modifying the schedule. However, this author feels that the planning and scheduling functions can be done automatically onboard the master spacecraft, using the master–slave architecture and appropriate techniques described above.

Some possible drawbacks of the master–slave architecture may be too much dependence upon the master (for failure reasons) and potential communications problems and large communications overhead needed to interact between the slaves and the master (and vice versa). Although these are possible limitations of this architecture, the advantages of such an approach (namely, achievement of global vs. local satisfaction, avoidance of temporal/spatial abstractions whereby a plan would not be consistent across the constellation while all spacecraft are updating it, and ease of development and implementation) seem to make this master–slave architecture worthy of consideration and further investigation.

References

AAAI Fall Symposium on Distributed Continual Planning (1998), Orlando, October.

Barrett, A. (1999), "Autonomy Architectures for a Constellation of Spacecraft," Proceedings of the International Symposium on AI, Robotics, and Automation in Space, the Netherlands, June.

Brann, D.B., D. Thurman, and C. Mitchell (1996), "Human Interaction with Lights-Out Automation: A Field Study," IEEE Conference Proceedings.

Chweh, C. (1998), "Autonomy in Space," *IEEE Intelligent Systems*, September/October.

Chien, S., B. Smith, G. Rabideau, N. Muscettola, and K. Rajan (1998), "Automated Planning and Scheduling for Goal-Based Autonomous Spacecraft," *IEEE Intelligent Systems*, September/October.

Chien, S., R. Knight, A. Stechert, R. Sherwood, and G. Rabideau (1999), "Using Iterative Repair to Increase the Responsiveness of Planning and Scheduling for Autonomous Spacecraft," IJCAI 99 Workshop on Scheduling and Planning Meet Real-Time Monitoring in a Dynamic and Uncertain World, Stockholm, August.

Cooke, D. and S. Hamilton (2000), "New Directions at NASA Ames Research Center," *IEEE Computer*, January.

Davydov, M. (1999), "Who Knows, Who Cares?," *Intelligent Enterprise*, Dec. 21.

desJardins, M. (2000), Interview with Marie desJardins, conducted by Jay Liebowitz, University of Maryland–Baltimore County, Catonsville, Maryland, February.

desJardins, M., E. Durfee, C. Ortiz, and M. Wolverton (1999), "A Survey of Research in Distributed, Continual Planning," *AI Magazine*, Vol. 20, No. 4.

Freuder, G. and M. Wallace, guest eds. (2000), "Constraints," *IEEE Intelligent Systems*, Special Issue, January/February.

Hexmoor, H. and S. Desiano (1999), "Autonomy Control Software," *The Knowledge Engineering Review*, Vol. 14, No. 4.

JMS North America Inc. (1995), Market Research and Program Report on Requirements for Building a Generic Expert Scheduling System, Fairfax, VA.

Liebowitz, J. (1999), "A Strategic Roadmap for Defining MIDEX Ground System Goals and Emerging Technologies," *Telecommunications and Space Journal*, Vol. 6.

Liebowitz, J. and W. Potter (1995), "Scheduling Objectives, Requirements, Resources, Constraints, and Processes: Implications for a Generic Expert Scheduling System Architecture and Toolkit," *Expert Systems with Applications Journal*, Vol. 9, No. 3.

LOGOS PSWG Working Group (1998), Draft Planning and Scheduling Report, NASA Goddard, September 8.

Multi-Agent Distributed and Collaborative Planning and Scheduling, http://www.cgi.com/web2/govt/collab.html.

Mudry, J. (2000), In-Space Internet Technologies for NASA Enterprises' Revolutionary Communications Concepts, NASA Glenn Research Center/Space Communications, http://spacecom.grc.nasa.gov/technologies/satellite/.

OpsSolver by Caleb Technologies Corp. (www.calebtech.com/OpsSolver.htm).

Pratt, S., R. Raines, C. Fossa, and M. Temple (1999), "An Operational and Performance Overview of the Iridium Low Earth Orbit Satellite System," *IEEE Communications Surveys*.

SaVi, University of Minnesota (www.geom.umn.edu/~worfolk/SaVi/).

Stoffel, A.W. and D. McLean (1997), "Tools for Automating Spacecraft Ground Systems: The Intelligent Command and Control Approach," *SpaceOps Proceedings*.

Tsang, E. (1995), "Scheduling Techniques: A Comparative Study," *British Telecom Technology Journal*, Vol. 13, No. 1.

Weng, M. and H. Ren (1997), "A Review of Multi-Agent Scheduling," University of South Florida, Tampa, FL, //www.eng.usf.edu/~ren/agentrev.html.

Wilkens, D. (1998), "Multi-Agent Planning Architecture," *Proceedings of the International Conference on AI Planning Systems*.

Wilkens, D. and M. desJardins (1999), "A Call for Knowledge-Based Planning," white paper, SRI International, AI Center, Menlo Park, CA.

Yunck, T., D. McCleese, W. Melbourne, and C. Thornton (1995), "Satellite Constellations for Atmospheric Sounding with GPS: A Revolution in Atmospheric and Ionospheric Research," Proceedings of the NASA Technology 2005 Conference, Chicago, October.

Zoch, D. and L. Hull (1991), "Scheduling Applications Interface Language (SAIL) Reference Manual," DSTL-91-021 Report, NASA Goddard, August.

Bibliography

Beck, J.C. and M. Fox (1998), "A Generic Framework for Constraint-Directed Search and Scheduling," *AI Magazine*, Vol. 19, No. 4.

Generic Spacecraft Analyst Assistant (GenSAA) and the Generic Inferential Executor (Genie) Promotional Booklet.

Liebowitz, J. (1993), "Methodology and Mapping between Problem Requirements and Solution Scheduling Approaches in Mission Planning Expert Scheduling Systems," NASA SBIR Phase I Final Report, AMEC/NASA Goddard.

NASA Planning and Scheduling Workshop (http://ic.arc.nasa.gov/ic/psworkshop), March 2000, San Francisco.

Planning/Scheduling Workshop at NASA Goddard (//sdcd.gsfc.nasa.gov/ISTO/SW/objectives.html).

Sabuncuoglu, I. (1998), "Scheduling with Neural Networks: A Review of the Literature and New Research Directions," *Production Planning and Control Journal*, Vol. 9, No. 1.

Special issue on "Distributed Continual Planning," *AI Magazine*, Vol. 20, No. 4, Winter 1999.

Index

9 780367 455316